# Data Analysis

## An Introduction

Jeffrey A. Witmer
*Oberlin College*

PRENTICE HALL, Englewood Cliffs, New Jersey, 07632

Editorial production/supervision: *Genevieve Lopez/Shirin Khan*
Pre-press Buyer: *Paula Massenaro*
Manufacturing Buyer: *Lori Bulwin*
Supplement acquisitions editor: *Susan Black*
Acquisitions Editor: *Steve Conmy*


A Simon & Schuster Company
Englewood Cliffs, New Jersey 07632



Printed in the United States of America

10 9 8 7 6 5 4 3 2 1

ISBN 0-13-205238-5

Prentice-Hall International (UK) Limited, *London*
Prentice-Hall of Australia Pty. Limited, *Sydney*
Prentice-Hall Canada Inc. *Toronto*
Prentice-Hall Hispanoamericana, S.A., *Mexico*
Prentice-Hall of India Private Limited, *New Delhi*
Prentice-Hall of Japan, Inc., *Tokyo*
Simon & Schuster Asia Pte. Ltd., *Singapore*
Editora Prentice-Hall do Brasil, Ltda., *Rio de Janeiro*

# Contents

**Preface** v

**Part I** **Single Variable Techniques**

**Chapter 1** **Graphing Data** 1

1.1 Stem-leaf diagrams 1
1.1.1 Outliers 3
1.1.2 Back-to-back stem-leaf diagrams 4
1.2 Dotplots and histograms 6
1.3 Boxplots 11
1.3.1 Side-by-side boxplots 13
1.4 Other graphing ideas 17

**Chapter 2** **Smoothing Data** 21

2.1 Median trace 21
2.2 Smoothing a time series 25
2.3 Complex smoothers 30

**Chapter 3** **Transformations** 32

3.1 Normal probability plots 32
3.2 Transformations 38

**Part II** **Dealing With Many Variables**

**Chapter 4** **Bivariate Relationships** 46

4.1 Scatter plots 46
4.2 Correlation 51
4.3 Resistant lines 57
4.4 Least-squares regression lines 62
4.4.1 $r^2$ 66
4.5 The regression effect 67

**Chapter 5** **Regression Diagnostics** 70

5.1 Residual plots 70
5.1.1 The Anscombe plots 74
5.1.2 Normal probability plots of residuals 76

5.2 Transformations in regression 77

5.3 Outliers and influential points 80

**Chapter 6** **Multiple regression** 84

6.1 Fitting multiple regression models 84

6.2 Interactions 92

**PART III** **Miscellaneous Topics**

**Chapter 7** **Collecting Data** 97

7.1 Surveys 97

7.2 Randomized response sampling 100

7.3 Observational studies vs. controlled experiments 103

7.4 Paired data 106

**Chapter 8** **Capture/Recapture** 108

8.1 Capture/Recapture 108

**Chapter 9** **Simpson's paradox** 110

9.1 Simpson's paradox 110

**Appendix** **Data Sets** 114

**Index** 121

# Preface

Most statistics textbooks present statistics as a collection of mathematical formulas applied to artificial problems. In recent years the computer has made it possible to use statistical graphics as an important aid in model fitting and in exploratory data analysis. The goal of this textbook supplement is to present the tools of data analysis in an engaging manner and at a level that is accessible to students in elementary classes. Most of the tools presented herein require little mathematical training. At the same time, these are techniques that are widely used by practicing statisticians.

Note that this book does not contain definitions and examples of basic statistical concepts such as the mean or the standard deviation - *Data Analysis: An Introduction* is meant to supplement, not replace, a standard statistics text.

OUTLINE

Chapter 1 covers elementary graphics for exploratory data analysis and data presentation. Chapter 2 demonstrates the use of smoothers for eliminating "noise" in a time series plot in order to reveal the underlying trends. Section 2 covers running medians of three, which are easily done by hand. Compound smoothers and the use of computers for smoothing are covered in Section 3.

Many statistical techniques are based on the assumption that the data at hand come from a normal distribution. Chapter 3 presents normal probability plots, which can be used to check this assumption, and then takes up the topic of transformations, for cases in which the data are not normal.

Chapter 4 introduces relationships between variables, while Chapter 5 presents the use of residual plots and other diagnostic tools in regression. Chapter 6 then extends ideas from the previous two chapters to the setting of multiple regression.

Chapter 7 focuses on the ways in which data are collected and on the difference between quality and quantity in data analysis. Chapters 8 and 9 present topics that can be presented at any time in a typical statistics course.

Parts I and III can be covered in almost any order. Some teachers will want to discuss topics in Part III early in their courses. Only Part II must be covered in order, as each chapter there builds on the previous one.

USING THE BOOK

I have used *Data Analysis: An Introduction* in two ways at Oberlin College. I teach a standard introductory (non-calculus based) course on statistical methods. Most texts at this level include some topics from exploratory data analysis, but I use *Data Analysis: An Introduction* to compliment such texts and to provide supplementary material and examples.

I also teach a course for advance undergraduates on mathematical statistics. Texts at this level rarely include real data and interesting applications. I supplement the theorems and proofs of mathematical statistics with the applications and data of *Data Analysis: An Introduction*. This seems to provide a good, balanced, view of statistics to those students who would not otherwise be likely to see the tools of applied statistics in their courses.

I expect that most students reading *Data Analysis: An Introduction* will have access to a statistics computer package and will use the computer extensively when analyzing data. Each package has its own special features, but most are helpful in applying the ideas presented in this book.

While writing I tried to keep in mind general themes and questions that help one understand data: Where did the data come from? What do these data reveal about the world around us? How can a graph help us see patterns and understand the data more clearly? I hope that the use of *Data Analysis: An Introduction* will make statistics courses more profitable and enjoyable, and that the book will help readers develop their abilities to analyze data and to understand the processes that generate data.

ACKNOWLEDGMENTS

Several persons critiqued early drafts and provided many helpful suggestions. In particular, I thank Murray Clayton, Roy Tamura, Jon Entin, and Sam Goldberg. I thank Steve Conmy, my editor, for his enthusiasm and guidance. I am indebted to the Alfred P. Sloan Foundation for a grant that supported my work on this book. As well, I am grateful to the members of the Statistics in the Liberal Arts Workshop for encouraging me during this project. Most of all, I thank Merrilee, Joel, and Kyle for their patience and support.

Jeffrey A. Witmer
Oberlin, Ohio

## 1.1 Stem-and-Leaf Diagrams

Here is a list of the seasonal snowfall totals in Cleveland, Ohio, for the winters 1970-71 through 1980-81 (the data have been rounded to the nearest inch)[1]:

51, 46, 69, 59, 67, 54, 63, 90, 38, 39, 61.

Each of these values is a two-digit number. The first number is 51, which has a tens digit of 5 and a ones digit of 1 (*51 = 5*10 + 1*). We can write this as 5 | 1. If we split each number into its tens digit and its ones digit we can easily display the data in a *stem-leaf diagram* in which the stems are the tens digits and the leaves are the ones digits. The stems are listed in order in a column to the left of a vertical line, with the leaves going on the right side of the line (there are usually several leaves on a single stem):

```
               3 | 8 9
               4 | 6
               5 | 1 9 4
     stems     6 | 9 7 3 1     leaves
               7 |
               8 |
               9 | 0
```

The numbers in the first row, 3 | 8 9, tell us that there are two values in the thirties: 38 and 39.

A stem-leaf diagram quickly shows the shape of the data. We can see that in a typical winter under study Cleveland got around 60 inches of snow, but that there was one unusual winter with 90 inches of snow. The stem-leaf diagram also allows the reader to reconstruct the original data set: not only do we know from the stem-leaf diagram that there is one value in the forties, we know that this value is precisely 46.

I constructed the stem-leaf diagram above as I read the data and, thus, the leaves are not in order. This is one of the nice features of stem-leaf diagrams: you can construct one as you read data for the first time. Nonetheless, it is convenient to clean up the diagram by arranging the leaves in order.

```
3 | 8 9
4 | 6
5 | 1 4 9
6 | 1 3 7 9
7 |
8 |
9 | 0
```

Here is another example. I often count the number of french fries that I am served, in a small bag, when I eat at a McDonald's restaurant. Here are my data. (You will have to trust that these numbers are accurate - my family and I have eaten the evidence.)

[1] From *Climatological Data, Ohio.*

```
3 | 4 5 7 9
4 | 0 3 5
5 | 2 3 5
```

Here 3|4 means that there were 34 french fries in a bag.

We can display these data a bit differently if we split the stems into parts. In the following stem-leaf diagram I split the values in the 30s into two parts: 30-34 and 35-39. I treated the 40s and 50s likewise.

```
3 | 4
3 |   5 7 9
4 | 0 3
4 |   5
5 | 2 3
5 |   5
```

As another example, I recorded the high school graduation rates for the 50 states and the District of Columbia in the following stem-leaf diagram[2]. I rounded each value to the nearest percent. In the diagram, 5|6 stands for 56%.

```
5 | 69
6 | 02334
6 | 556777889
7 | 001223334444
7 | 56677888999
8 | 01123
8 | 566789
9 | 1
```

Note that this distribution is fairly symmetric and bell-shaped, with no values that stand out from the others.

Another possible feature in a stem-leaf diagram is a *depth column*, which helps to locate the center of the data:

```
depth
   2  5 | 69
   7  6 | 02334
  16  6 | 556777889
(12)  7 | 001223334444
  23  7 | 56677888999
  12  8 | 01123
   7  8 | 566789
   1  9 | 1
```

2 These are public high school graduation rates for 1987, as reported by the U.S. Department of Education.

The depth column on the left tells us how far we have come as we move from an edge of the diagram in toward the middle of the data set. There are 2 values in the 50's, 7 values that are less than 65, and so on. If one person counts from the top down and another person counts from the bottom up, they will meet in the 70-74 stem. Thus, the *median* is in the 70-74 group. The median is the 50th percentile, which splits the data in half - half of the values are less than or equal to the median and half of the values are greater than or equal to the median.

There are 12 values in the central group, 16 smaller values, and 23 larger values. Hence, there are 51 (12 + 16 + 23) values in this data set. The median is the 26th-largest observation: 74.

If the number of observations is even, rather than odd as in the last example, then there is no single observation that splits the data into a top half and a bottom half. In this case, we define the median to be the average of the two observations that come closest to being directly in the middle. For example, the winning teams in the college football bowl games played on New Year's Day, 1991, scored a very large number of points. Here are the point totals of the winning teams, sorted from smallest to largest:

10, 23, 30, 34, 35, 45, 46, 46

In this case, the 4th largest and 5th largest observations tie for the "honor" of being closest to the middle. To find the median we average these two values: (34 + 35)/2 = 34.5.

### 1.1.1 Outliers

The 1980 census found that the median age in the U.S. population was 30 years. The median age varies from state to state; for example, Wisconsin had a median age of 29.4 years. The following stem-leaf diagram shows the median ages in the 50 states. The stems are years and the leaves are tenths of years.

```
   1  24 | 2
      25 |
   2  26 | 1
   7  27 | 14467
  15  28 | 12346789
(16)  29 | 0122234467888999
  19  30 | 011112234469
   7  31 | 289
   4  32 | 012
      33 |
   1  34 | 7
```

Notice that there are two numbers that stand out from the others. The value 34.7 belongs to Florida and the value 24.2 is for Utah. These two observations are called *outliers*. They are so far from the other observations that they warrant special consideration. When you encounter an outlier you should ask yourself such questions as “Where did this value come from? Is this the result of a typographical error? Of some other kind of error? Is there a special explanation for the outlier?” In this case, it comes as no surprise that Florida had a high median age - it has a high percentage of senior citizens. On the other hand, you may be surprised that Utah had such a small median age, but this is not a typographical or recording error.

### 1.1.2 Back-to-back stem-leaf diagrams

When comparing two sets of data it is often helpful to construct back-to-back stem leaf diagrams. For example, consider the two most productive home run hitters in baseball history, Babe Ruth and Hank Aaron. The following stem-leaf diagrams show the yearly home run totals of each player for years in which the player had at least 350 at-bats. Here, 6 | 0 stands for 60 home runs.

```
      Ruth

  3   2 | 259
  5   3 | 45
 (7)  4 | 1166679
  4   5 | 449
  1   6 | 0
```

```
      Aaron

  2   1 | 23
  6   2 | 4679
 (7)  3 | 0244899
  8   4 | 00444457
```

These distributions are easier to compare if we put the diagrams back-to-back:

```
      Ruth        Aaron

            | 1 | 23
        952 | 2 | 4679
         54 | 3 | 0244899
    9766611 | 4 | 00444457
        944 | 5 |
          0 | 6 |
```

This display clearly shows that Aaron's best seasons were equivalent to average seasons for Ruth. The differences between the two are even more striking if we split the stems.

```
      Ruth        Aaron
            | 1 | 23
            | 1 |
          2 | 2 | 4
         95 | 2 | 679
          4 | 3 | 0244
          5 | 3 | 899
         11 | 4 | 004444
      97666 | 4 | 57
         44 | 5 |
          9 | 5 |
          0 | 6 |
```

Exercises

1. Below is a stem-leaf diagram for seasonal snowfall in Cleveland, OH, for the winters 1939-40 through 1981-82.

```
 3 | 111677778899
 4 | 356689
 5 | 0222334445669
 6 | 1134799
 7 | 567
 8 |
 9 | 0
10 | 1
```

(a) Construct a depth column for this diagram and find the median.
(b) Write a sentence that describes the distribution.

2. Here is a list of 13 western states and the 1989 divorce rate for each state.

| | | | | | |
|---|---|---|---|---|---|
| Alaska | 6.3 | Idaho | 6.0 | Oregon | 5.4 |
| Arizona | 6.5 | Montana | 5.1 | Utah | 4.7 |
| California | 4.3 | Nevada | 11.9 | Washington | 5.7 |
| Colorado | 5.6 | New Mexico | 5.0 | Wyoming | 6.4 |
| Hawaii | 5.1 | | | | |

(a) Make a stem-leaf diagram of these data, in which the digits to the left of the decimal point are the stems and digits to the right of the decimal point are the leaves (i.e., so that 6|3 represents 6.3). Include a depth column for your stem-leaf diagram.
(b) Write a sentence that describes the distribution.

3. Here are the yearly batting averages of Pete Rose for seasons in which he had at least 350 at-bats: .273, .269, .312, .313, .301, .335, .348, .316, .304, .307, .338, .284, .317, .323, .311, .302, .331, .282, .325, .271, .245, .286, .264.
Here is the corresponding list for Ty Cobb: .320, .350, .324, .377, .385, .420, .410, .390, .368, .369, .371, .383, .382, .384, .334, .389, .401, .340, .338, .378, .357, .323.
Construct back-to-back stem-leaf diagrams of these data. Compare the yearly batting averages of Rose and Cobb.

4. The following stem-leaf diagram shows the ages at inauguration of the Presidents of the United States. In this diagram, the 40's were split into two parts: 40-44 and 45-49. Thus, there are two stems for the 40's, likewise for the 50's and the 60's.

```
4 | 23
4 | 67899
5 | 001111224444
5 | 555566677778
6 | 011124
6 | 4589
```

(a) Construct a depth column for this diagram.
(b) Find the median age and comment on the distribution.

## 1.2 Dotplots and Histograms

Many statistical computer packages generate *dotplots* (also known as *lineplots*), which are useful when working with a moderate number of points. A dotplot gives one a quick feel for the distribution of the data. Here is a dotplot of the high school graduation rates in the U.S.[1]

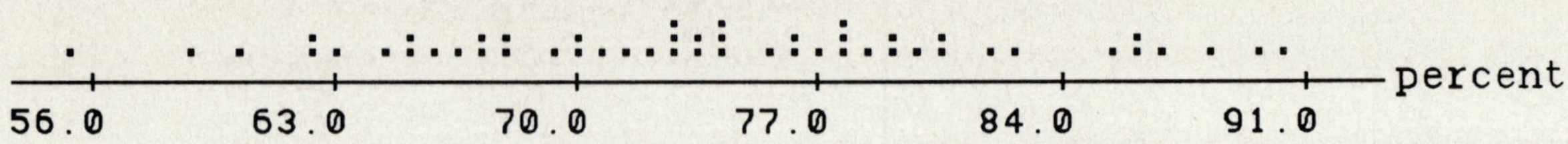


Each dot represents the graduation rate for a state. The range 56-91 is divided into several intervals. Each observation is plotted above the interval in which it falls. We can compare states east of the Mississippi River to those west of the Mississippi by producing separate dotplots.

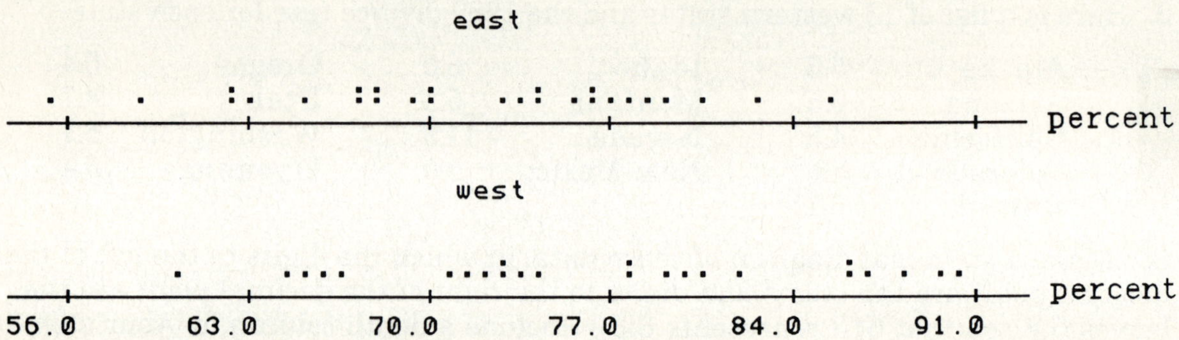


We can see that the graduation rates in western states are shifted slightly above those in the east.

Another means of comparing groups is to construct a dotplot that uses different plotting symbols for each group. For example, here are more counts of french fries per small bag at three fast-food restaurants.

McDonald's: 35, 39, 43, 52, 53, 55, 34, 37, 45, 40
Burger King: 51, 55, 52, 43, 44
Rax: 41, 48, 43, 57, 41, 43, 60, 51, 49, 55, 43, 48, 40, 48

In the following dotplot, McDonald's values are plotted with the symbol **m**, Burger King with the symbol **b**, and Rax with the symbol **r**.

---

1 As with many other data sets presented in this book, the raw data may be found in the Appendix.

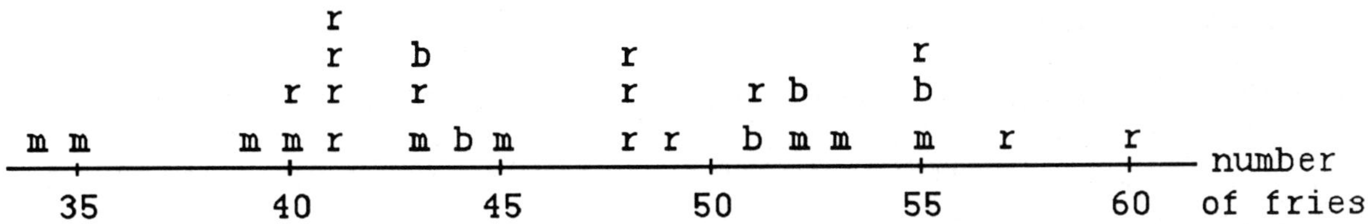


As the number of observations to be plotted grows, it becomes more and more difficult to construct a stem-leaf diagram or a dotplot. We can display a large set of data with a *histogram*, which gives the shape of a distribution by showing number of observations in each of several classes or intervals. For example consider the 1989 salaries of 318 National League baseball players. To make a histogram we must divide the data into several classes and count the number of observations in each class. (This is most easily done with a computer!) Using the classes

$$\text{salary} \leq \$200{,}000$$
$$\$200{,}000 < \text{salary} \leq \$400{,}000$$
$$\vdots$$

yields the following histogram.

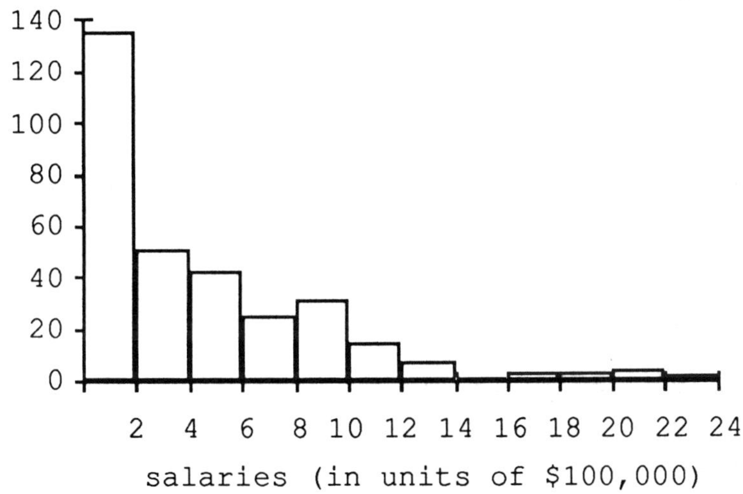


Note that this distribution is highly skewed, having a long right-hand tail.

We can display the baseball salary data with a slightly different histogram if we combine all of the salaries above $800,000 into one big category.

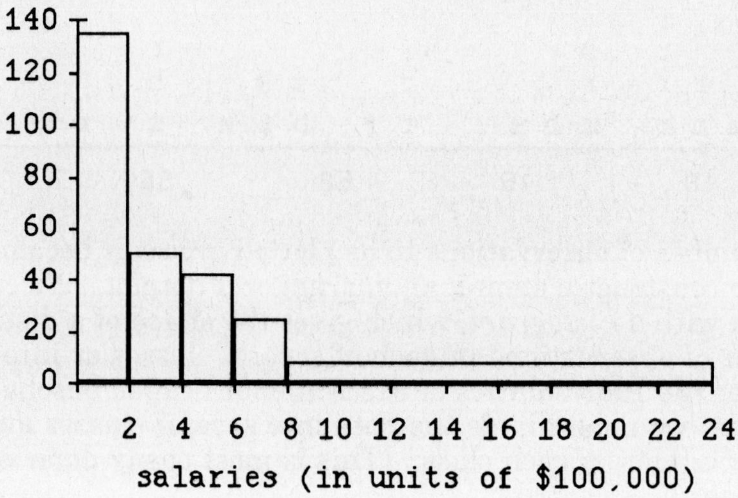


Since the final interval above is 8 times as long as the other intervals, we make the final rectangle 1/8th as high as we would otherwise, so that the area of each rectangle is proportional to the number of observations it represents. The histogram represents the number of observations in each class with rectangles - the more observations, the larger the rectangle. But it is the amount of area in each rectangle that conveys to the observer how many observations are in a class.

Notice how the following incorrect histogram gives a distorted impression of the data. (Here the final interval is 8 times too tall.)

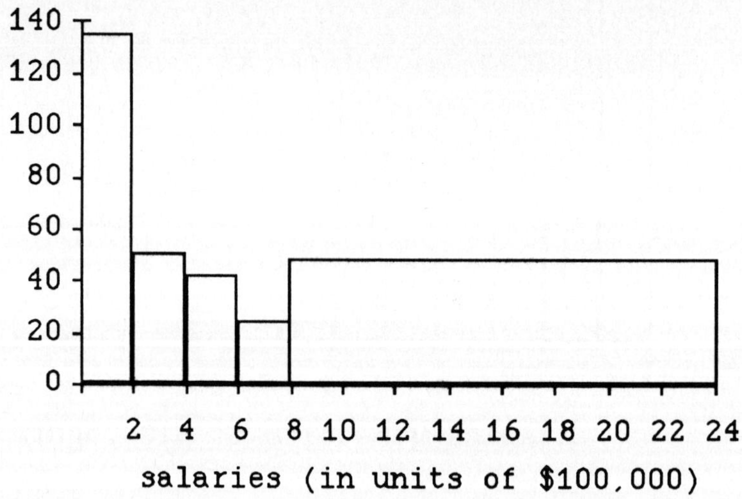

Exercises

1. Here are data on the percentage of space in each of several magazines that is taken up by advertisements. (These data were collected by Sandra Spyke, Marilyn Stano, and Dan Mussulin in July, 1990.)

| Magazine | % Advertising | Magazine | % Advertising |
|---|---|---|---|
| *Business** | 20 | *Nat'l. Geog.* | 15 |
| *Byte** | 59 | *New Yorker* | 27 |
| *CD-ROM Review** | 52 | *Newsweek* | 30 |
| *Discover** | 31 | *P C Publishing** | 29 |
| *Ebony* | 41 | *Sports Ills.* | 43 |
| *Industry Week** | 45 | *Time* | 35 |
| *Life** | 38 | *Today's Office** | 36 |
| *Macleans* | 33 | *US News & World Rpt.* | 39 |
| *Money** | 60 | *Wasn. Jour. Rev.** | 27 |
| *Math. Teacher* | 5 | *World Press** | 32 |

(a) Construct a dotplot of these data. Note: The magazines that cost over $2.75 are marked with asterisks. You might want to compare these to the less expensive magazines, for example, with separate dotplots.
(b) Construct a histogram of these data.
(c) Describe the distribution.

2. Consider the 1988 salaries of 321 NBA basketball players. These range from $100,000 (earned by several players) to a high of $3,142,860 (for Magic Johnson of the Lakers). Using the classes

$$\$100,000 \leq \text{salary} \leq \$300,000$$
$$\$300,000 < \text{salary} \leq \$500,000$$
$$\vdots$$

yields the following histogram.

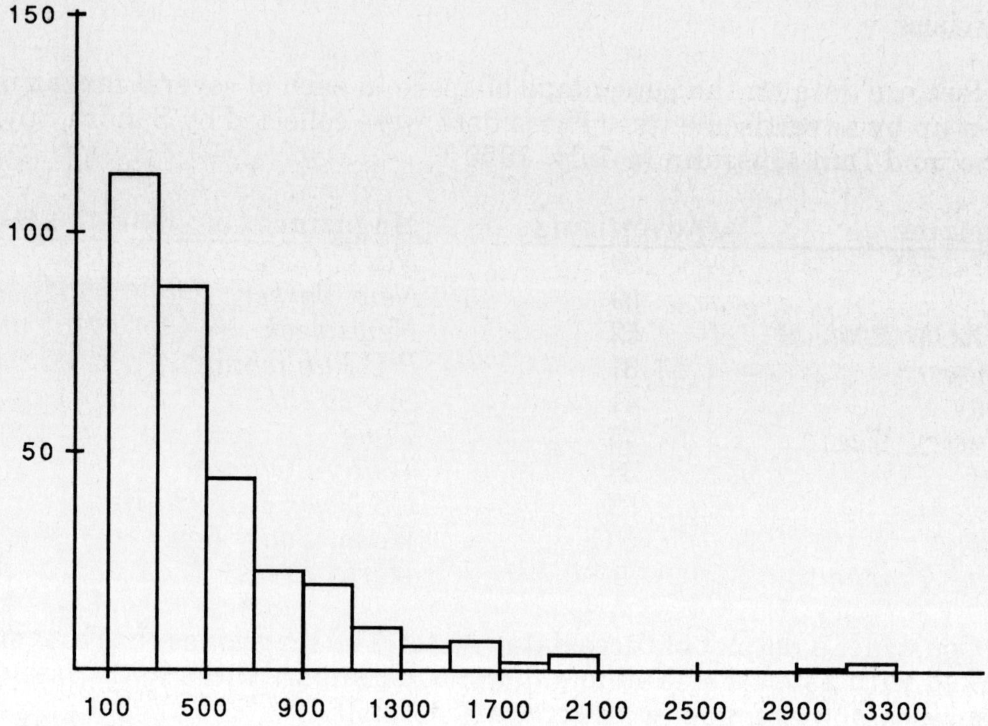


Describe this distribution.

## 1.3 Boxplots

It is convenient to summarize a set of data by reporting five numbers - the maximum, the minimum, the median, the first quartile (denoted by $Q_1$), and the third quartile (denoted by $Q_3$). The first quartile is the 25th percentile, the value that cuts off the bottom quarter of the data. The third quartile is the 75th percentile, the value that cuts off the top quarter of the data. Consider the high school graduation rate data, which are presented in order, from smallest to largest value, in the following stem-leaf diagram.

| | | |
|---|---|---|
| 2 | 5 | 69 |
| 7 | 6 | 02334 |
| 16 | 6 | 556777889 |
| (12) | 7 | 001223334444 |
| 23 | 7 | 56677888999 |
| 12 | 8 | 01123 |
| 7 | 8 | 566789 |
| 1 | 9 | 1 |

This data set has 51 observations, with a median of 74, which is the 26th-largest value. There are 25 values less than the median. The first quartile splits these 25 values into an upper half (of 12) and a lower half (of 12). Thus, the first a quartile is the 13th-largest observation, which is 67. The third quartile is found in a similar manner. Here, $Q_3$ is the 38th-largest observation, which is 79[1]. Thus, we have

12 values, $Q_1$, 12 values, median, 12 values, $Q_3$, 12 values

If we remove the District of Columbia from the data set there are 50, rather than 51, observations. In this case, the median is the average of the 25th largest and 26th largest observations. There are still 25 values larger than the median (we include the 26th largest observation in this group) and 25 values less than the median, so the quartiles do not change. If there were 48 observations in the data set then the median would be the average of the 24th and 25th largest values. We would determine the first quartile by finding the middle of the smallest 24 values. In this case $Q_1$ would equal the average of the 12th largest and 13th largest observations.

A statistical graphic that presents the five numbers - minimum, $Q_1$, median, $Q_3$, maximum - is the *boxplot* or box-and-whisker diagram. For the high school graduation rate data we have

| | | |
|---|---|---|
| minimum | = | 56 |
| $Q_1$ | = | 67 |
| median | = | 74 |
| $Q_3$ | = | 79 |
| maximum | = | 91 |

[1] Note: Some authors give slightly different definitions for the first and third quartiles.

The following boxplot displays these five values.

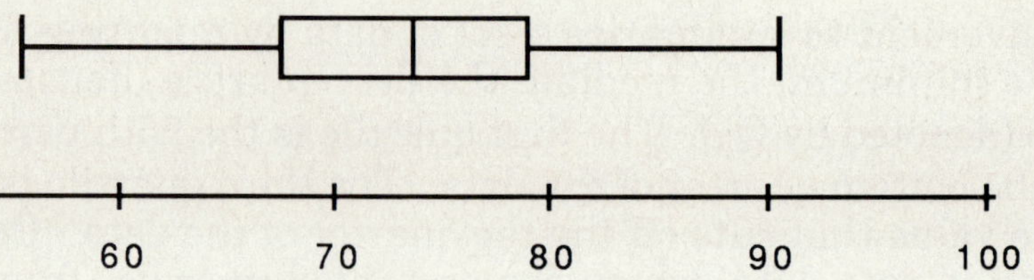


The box in the plot extends from the lower quartile to the upper quartile. The lines (whiskers) extend down to the minimum and up to the maximum. The median is indicated by the line in the middle of the box and locates the center of the distribution.

The boxplot also shows us how variable the data are. Two measures of variability, or spread, are the *range* and the *interquartile range*. The range is given by

$$\text{range} = \text{maximum} - \text{minimum}.$$

The interquartile range or *IQR* is given by

$$IQR = Q_3 - Q_1$$

and equals the length of the box. Of the two, the IQR is a better measure of the variability of the data since the IQR is not affected by a few extreme values. That is, the IQR is robust.

Notice that the boxplot gives a quick visual summary of the data set. We see immediately the center of the data and the variability of the data, as well as the maximum and minimum. Unlike a stem-leaf diagram, a boxplot is useful for displaying very large sets of data.

If there are outliers in the data set then a slight modification is needed in the boxplot. An outlier is an observation that is far removed from the bulk of the data. We take here the following definition of outlier - *an observation is an outlier if it lies more than 1.5*IQR from the end of the box*. We expect the largest few observations to be a fair distance above the upper quartile, but if any of them exceed $Q_3$ by more than 1.5 times the length of box we label them as outliers. Likewise for small observations - if they are more than 1.5*IQR below $Q_1$ we call them outliers.

*Warning: outliers are not to be ignored; they are often the most interesting values in a data set. We should always try to account for them before proceeding with the data analysis.*

Outliers are shown on boxplots by being plotted as separate points beyond the whiskers. If a data set contains one or more outliers we extend the whiskers of the boxplot from the end of the box to the last point that is not an outlier. We then plot the outliers as separate points. For example, in the statistical software package *Data Desk*® the symbol o is used to plot outliers that are within 3*IQR of the end of the box and the symbol * is used to plot outliers that are more then

---

Data Desk is a registered trademark of Data Description, Inc.

3*IQR from the end of the box. (*Data Desk* also puts the ends of the box at the hinges, which are slightly different from the quartiles.)

The following boxplot was produced using *Data Desk* and displays the distribution over the years 1938-1978 of June rainfall, in inches, in Cleveland, OH. Note that there are two outliers, one of which (the maximum) is more than 3*IQR above $Q_3$.

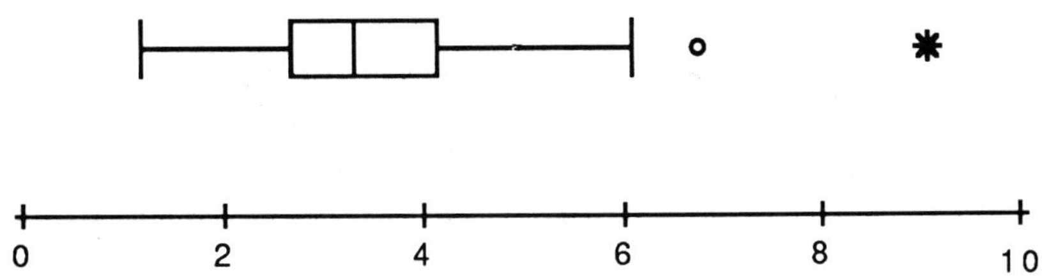


1.3.1 Side-by-side boxplots

An excellent way to compare distributions quickly is to construct side-by-side boxplots. The following boxplots are for the heights of a sample of college students, broken down by sex. Looking at the boxplots gives one a quick sense of the extent to which the two distributions overlap and the ways in which they differ. Notice that the lower quartile of heights of the men is just a bit above the upper quartile for the women. The tallest woman, who is an outlier with respect to the other women, is at the upper quartile for the men.

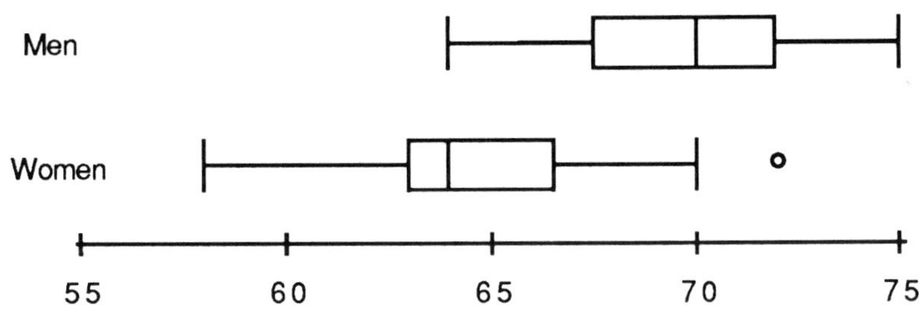


As another example, the following are boxplots of the state-by-state percentage of the popular vote that George Bush received in the 1988 Presidential election, broken down by region. In the south the median overlaps the upper edge of the box. Note that the minimum in the south is above the median for the east or the midwest.

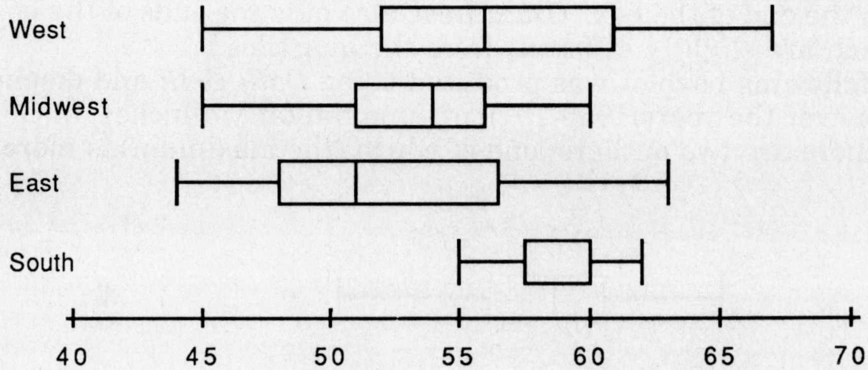


Here is a boxplot of the daily high temperature in Oberlin, OH for each day in 1988.

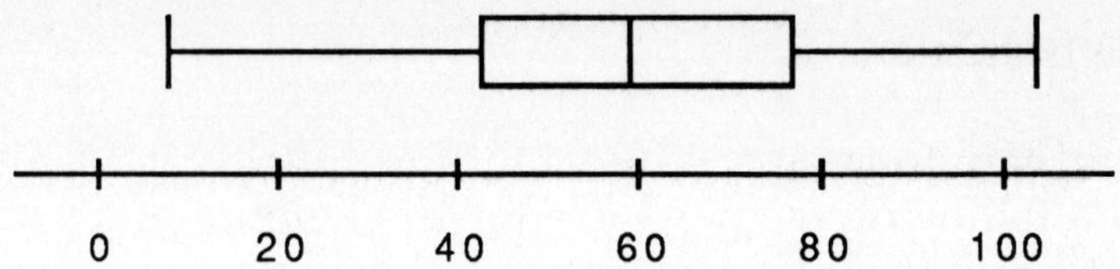


The plot shows that on a typical day the high temperature was near 60 degrees, with the high being between 40 and 80 degrees more than half of the time. The hottest day saw the temperature rise slightly above 100 degrees, and the lowest high temperature was near 10 degrees.

Of course, there is quite a bit of change in the weather from one month to another. Here are side-by-side boxplots for each month.

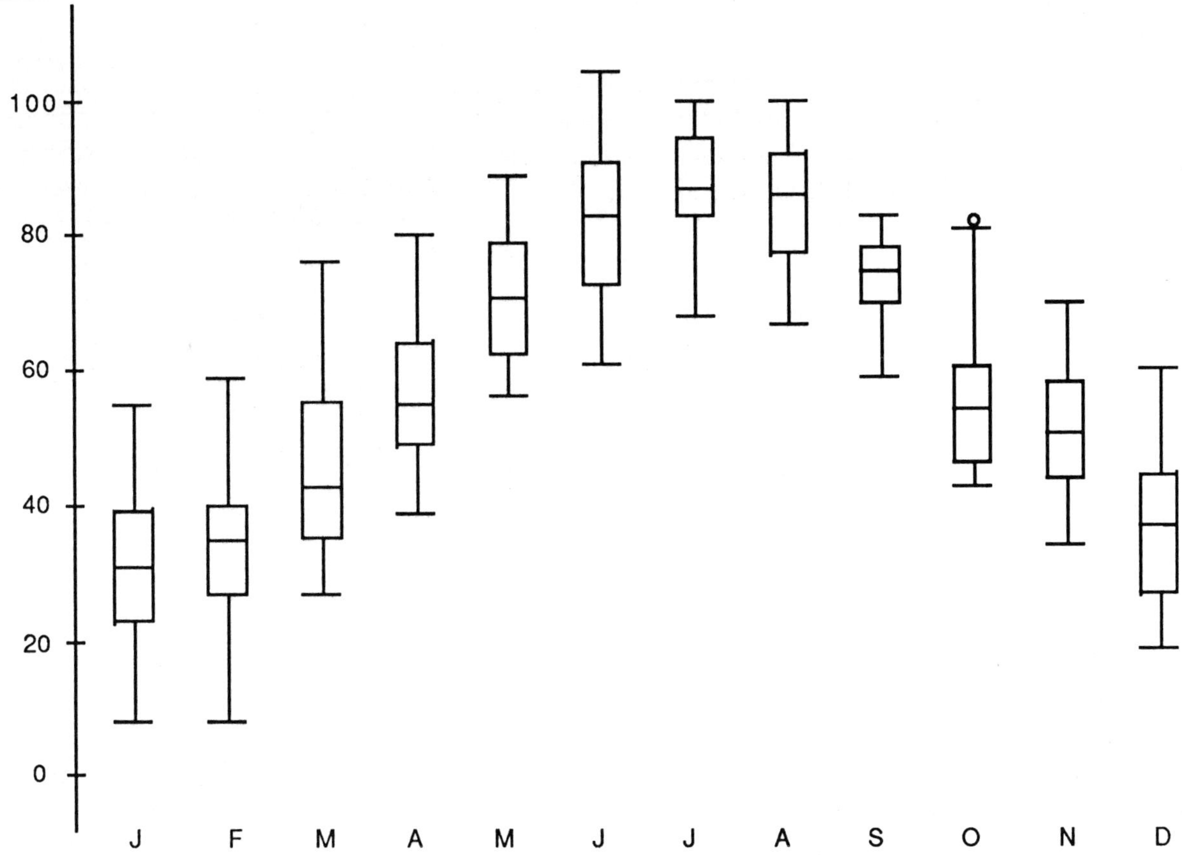


These plots allow us to quickly compare months and to see how the high temperature varies as the year progresses. Note that there is slightly more variability in the winter months than in the summer, as indicated by the lengths of the boxes and the whiskers.

## Exercises

1. The Los Angeles Dodgers often lead the National League in home attendance. The following are boxplots for attendance at home baseball games of the Dodgers for the first half of the 1989 season. One boxplot is for weekday games (Monday - Thursday), while the other is for weekend games. Comment on each of these distributions (What is the shape of the distribution? Is it symmetric? How large is the median? Are there any outliers? etc.) and compare them.

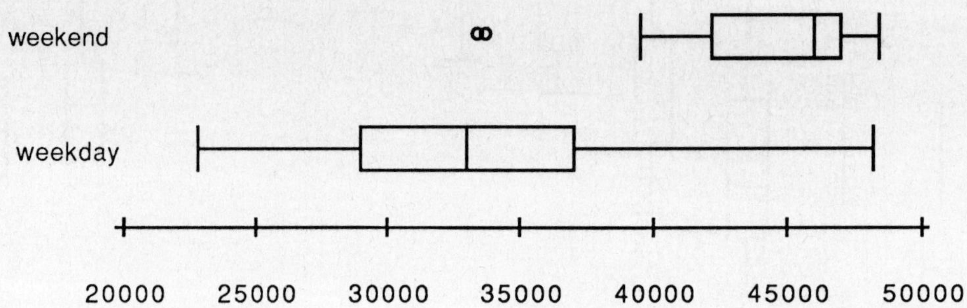


2. The following boxplots show state-to-state variability in the percentage of the popular vote that the winning candidate received in the 1976, 1980, 1984, and 1988 U.S. Presidential elections. Comment on these plots. (Utah is an outlier on the low side in 1976, when Jimmy Carter won, and on the high side in 1980 and 1984, when Ronald Reagan won. The two high outliers in 1976 are Arkansas and Georgia.)

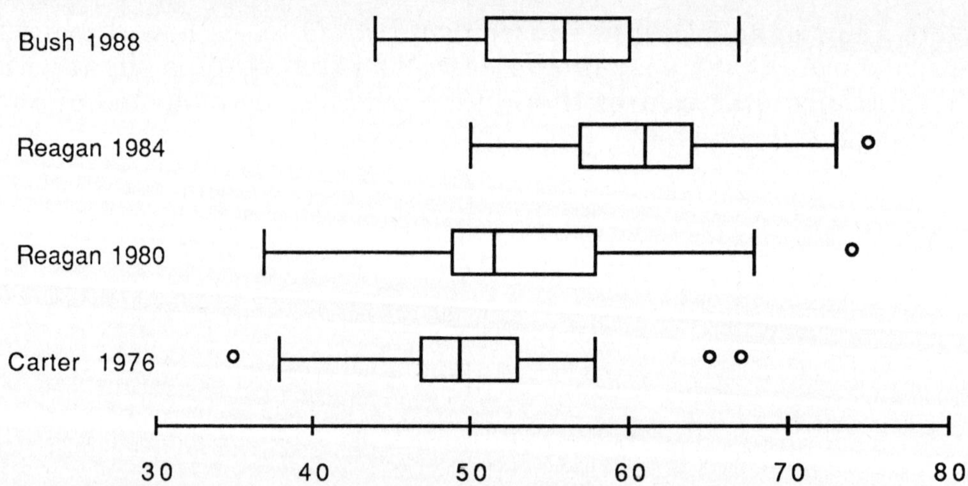


3. Consider the divorce rate data from exercise 2 in Section 1.1.
(a) Find the median, the first quartile, $Q_1$, and the third quartile, $Q_3$.

(b) Use the 1.5*IQR criterion to determine if there are any outliers. If there are any outliers, which states are they?

(c) Construct a boxplot of the data.

## 1.4 Other Graphing Ideas

There are many others ways to present data. One of the most common graphics is the *bar chart*. The following bar chart shows the color distribution for a random package of M&Ms.

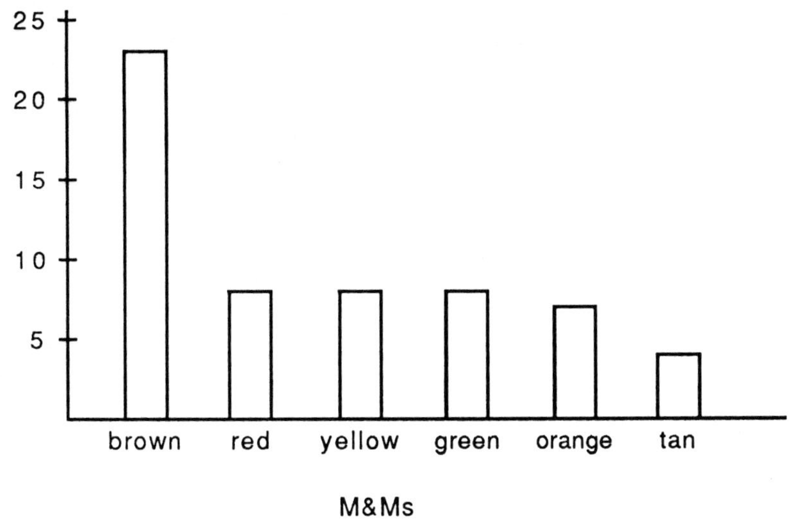


People sometimes confuse bar charts with histograms. A bar chart is used to display observations that fall into nominal categories, such as red, yellow, and green above. We can rearrange the categories in the M&M bar chart if we wish:

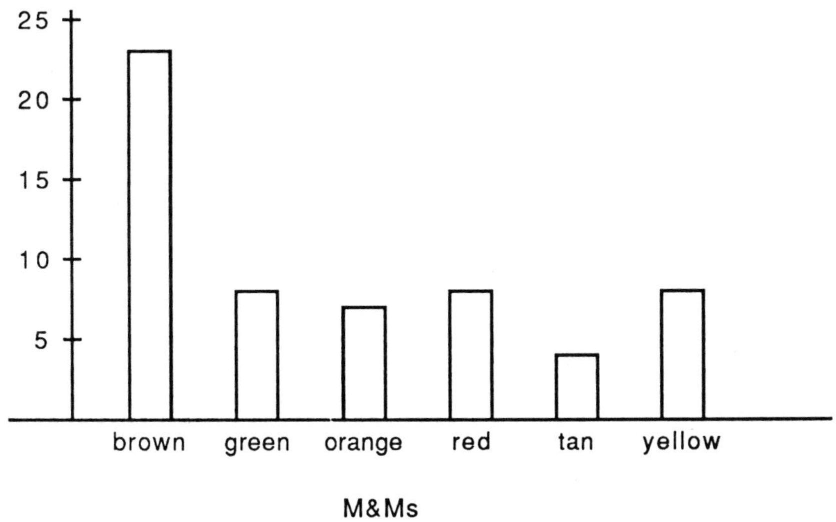


A histogram, on the other hand, is used to display numerical data, such as income, age, or temperature. We create the groups for a histogram by dividing the number line into intervals. We cannot then arbitrarily move an interval from the left side of a histogram to the right side, for instance.

A common but largely useless graphic is the *pie chart*. Here is a pie chart that shows the racial/ethnic breakdown of bachelor's degree recipients at U.S. colleges in 1987. Each slice of the pie represents a racial/ethnic group, with the size of the slice proportional to the number of degrees awarded in that group.

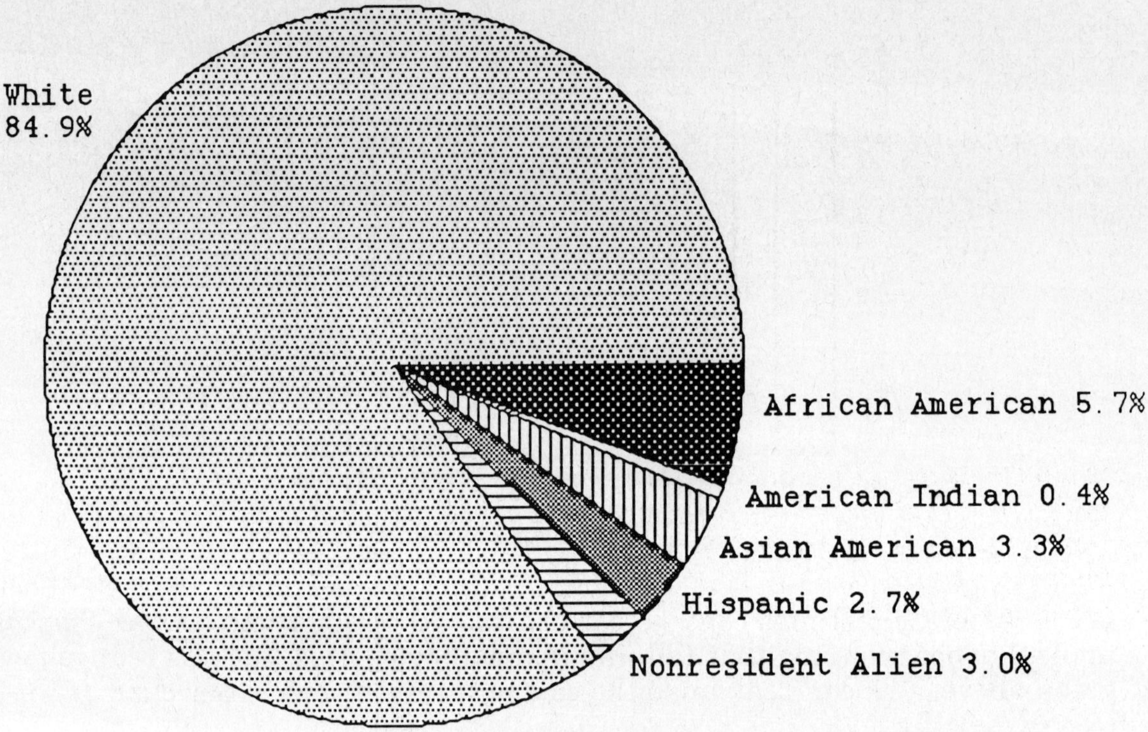


The problem with this graphic is that it is difficult to judge the size of a small slice of the pie and, so, we resort to reading the percentages given, which largely defeats the purpose of using a graphic!

Often two pie charts are juxtaposed for comparison purposes. The following pie charts show the 1980 racial distributions in California and New York.

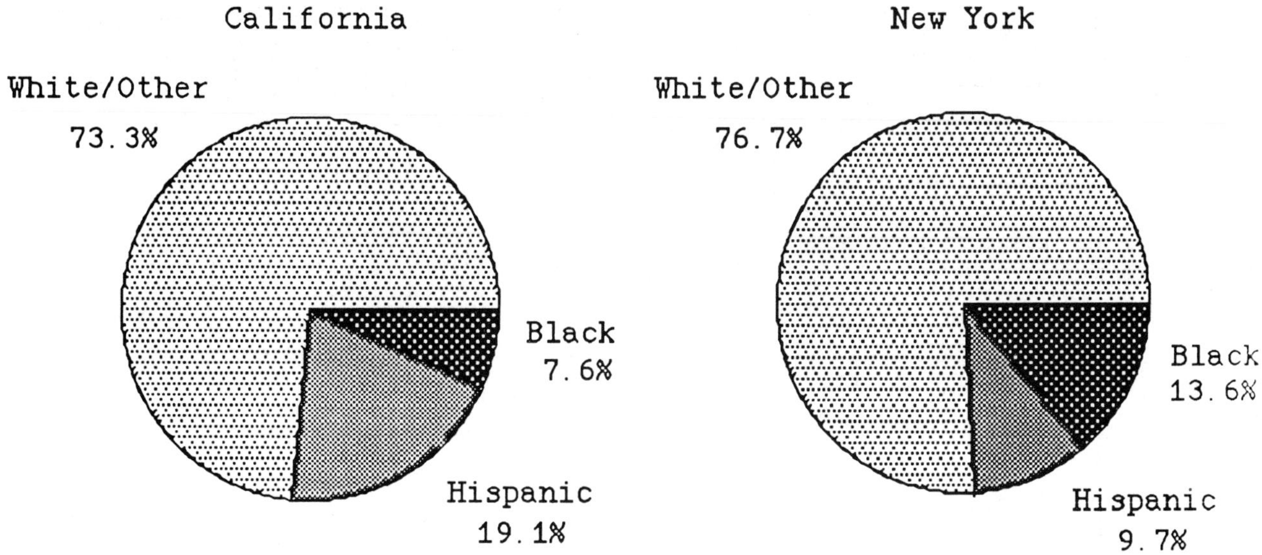


It is difficult to visually compare two pieces of pie, so the reader generally ends up reading the percentages associated with the slices. Pie-charts are sometimes useful for presenting the division of a whole (the pie) into a few moderately sized parts (the slices). But when the number of parts gets large, some of the slices are thin, or we want to compare two distributions, it is best to avoid pie-charts.

Often data are collected over time and are displayed in a *line chart*, such as the following.

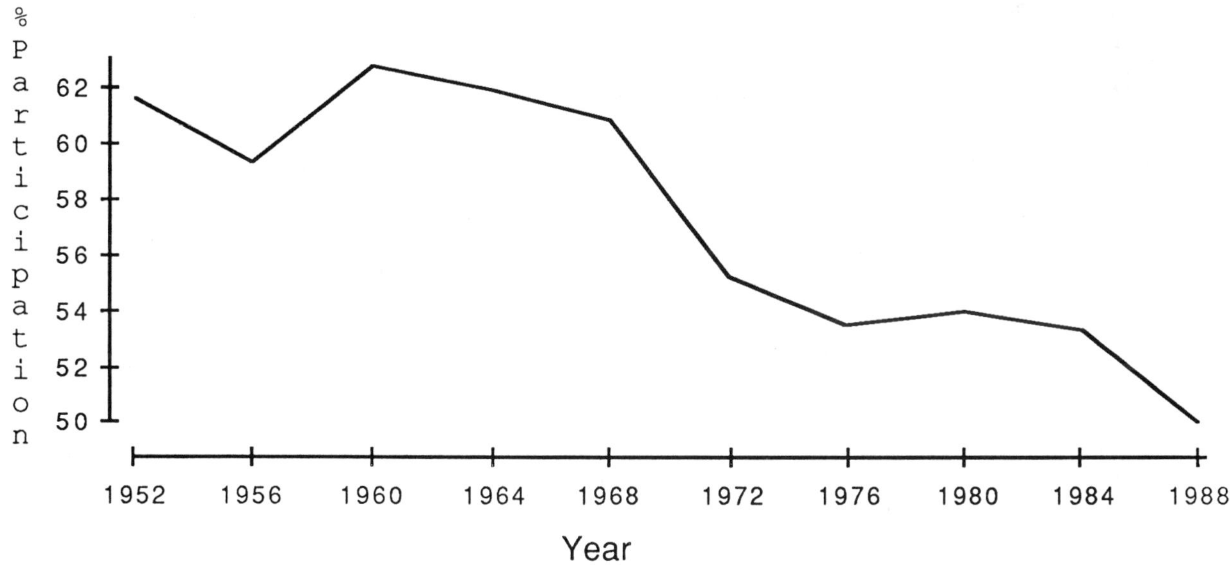


By following the line we see the trend over time in voter participation in Presidential elections. Such a graph is easy to construct and to read. When reading such a graph, be sure to check that the values are correctly spaced on the horizontal axis. Notice the effect of improper spacing.

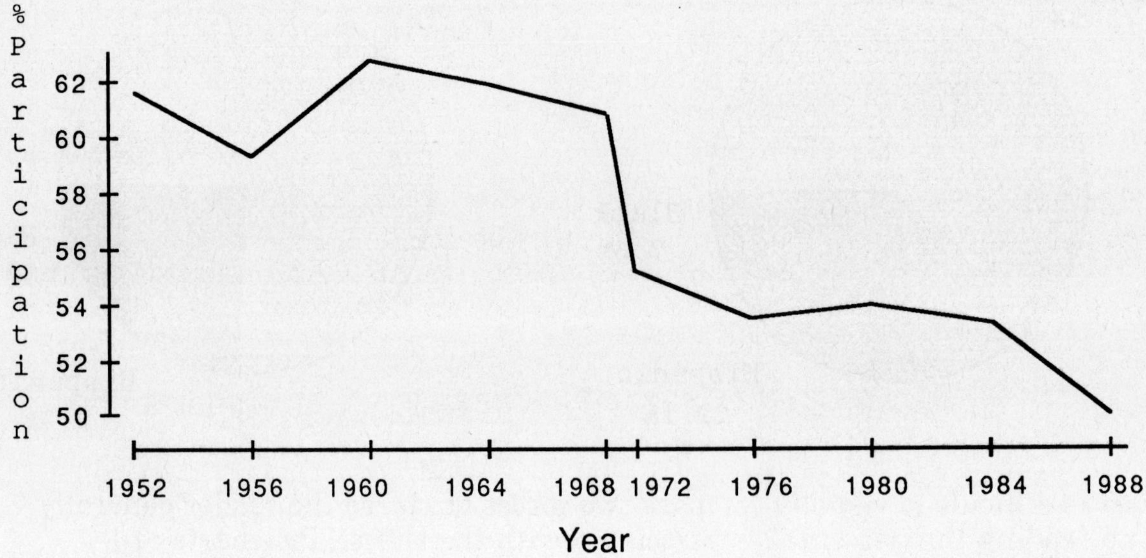


In this line chart the years are not evenly spaced, which makes the change between 1968 and 1972 seem even more dramatic.

Much has been written about sophisticated graphing. An excellent reference is Edward Tufte's book *The Visual Display of Quantitative Information*[1].

---

1 E.R. Tufte (1983) *The Visual Display of Quantitative Information*, Cheshire, Connecticut: Graphics Press.

2.1 Median trace

The 1970 draft lottery was an attempt by the Selective Service to draft men at random. The idea was that a selection number between 1 and 366 would be assigned at random to each possible birthday, with men born on days having small selection numbers being drafted first. It turned out that selection number 1 was assigned to September 14th, which is the 258th day of the year, so that men born on September 14th were drafted first; selection number 366 was assigned to June 8th (which is day number 160) so that men born June 8th would have been drafted only after all other men were drafted, i.e., having a large selection number is good (assuming you do not want to be drafted).

The following is a plot of selection number vs. birthday[1]. If things really were random then there would be no pattern in this plot.

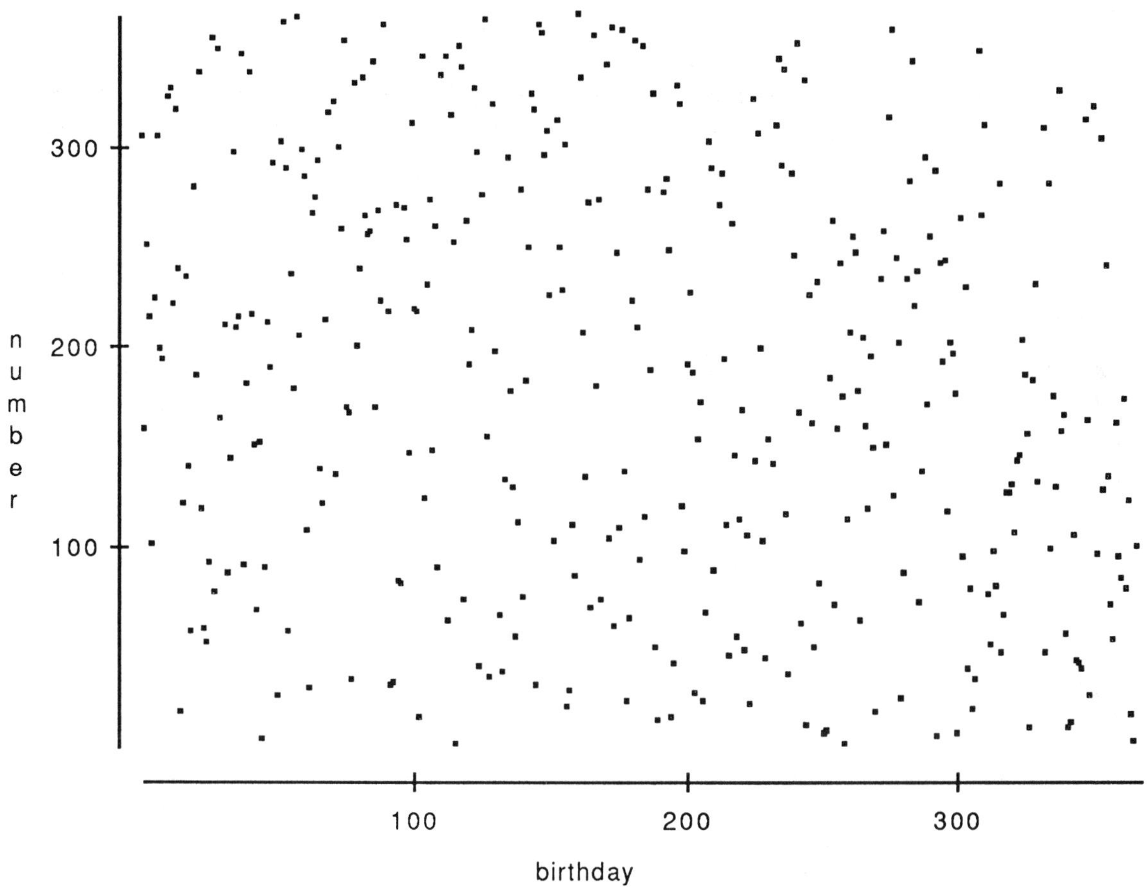


There is a trend in this plot, although it is not obvious. When the lottery was conducted the 31 capsules for January were first placed in a large box. Then the 29 February capsules were added and were mixed with the January capsules; the 31 March capsules were added and mixed in next, and so on. The December capsules were added last.

---

[1] For an extensive discussion of the draft lottery see S. E. Fienberg (1971) "Randomization and social affairs: the 1970 draft lottery," *Science* **171**:255-261.

When the capsules were drawn out those for December tended to be drawn first, giving low draft numbers to men born in December. To see this, consider the following *median trace*. This statistical graphic connects the median draft numbers for the 12 months. To construct a median trace, we first slice the scatter plot, like the side view of a loaf of bread, into vertical strips. In this case, it is natural to make the strips correspond to the 12 months of the year. Within each strip we find the median draft number. For example, the median of the 31 draft numbers assigned to December birthdays was 100, whereas the median draft number for January was 211. We then plot the median draft numbers on the original scatter plot and connect them with lines, so that we can see the pattern in the medians:

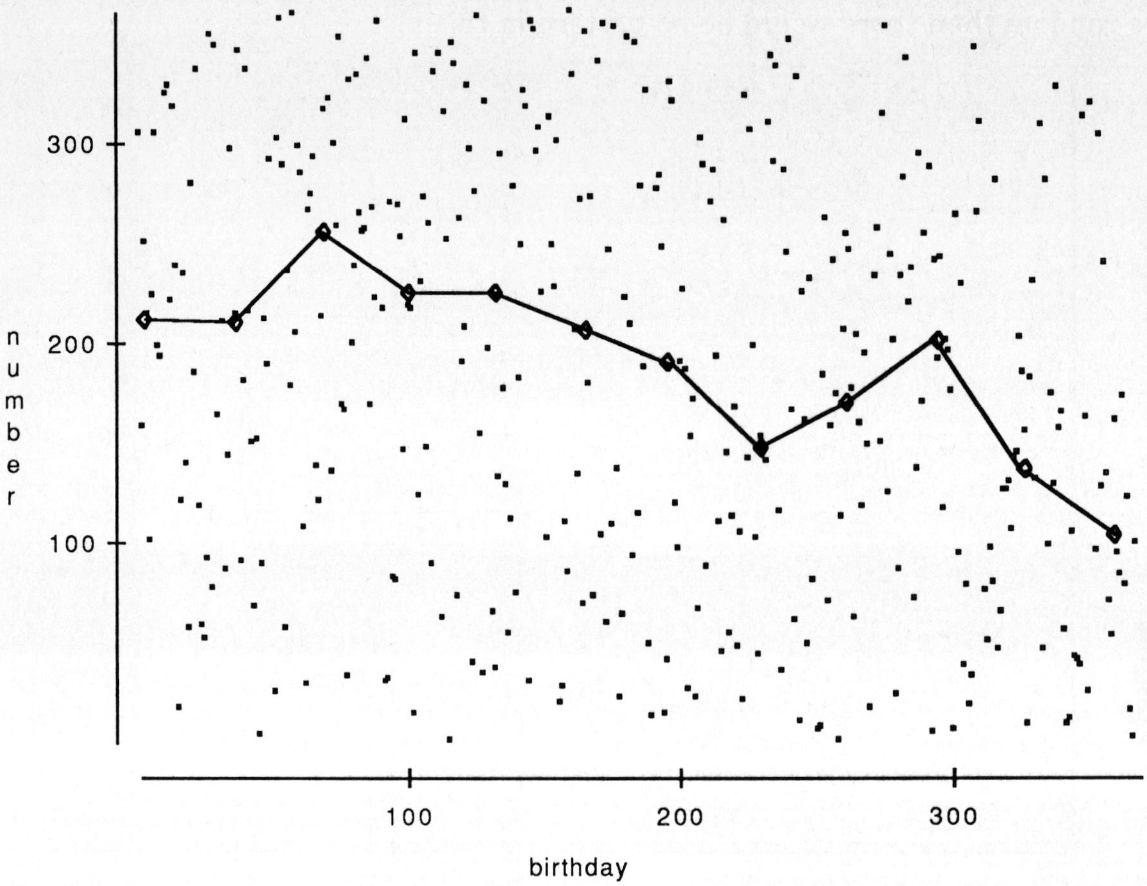


We noted earlier that the process by which the capsules were added and mixed in the box resulted in the January capsules being mixed several times, while the December capsules were mixed with the others only once. Thus, in addition to having a small median, we would expect the December draft numbers to be less variable than those from other months. We can investigate the variability <u>and</u> the median for each month by looking at side-by-side boxplots. Note the the interquartile range (the length of the box) for December is small when compared to the other months, as we expected.

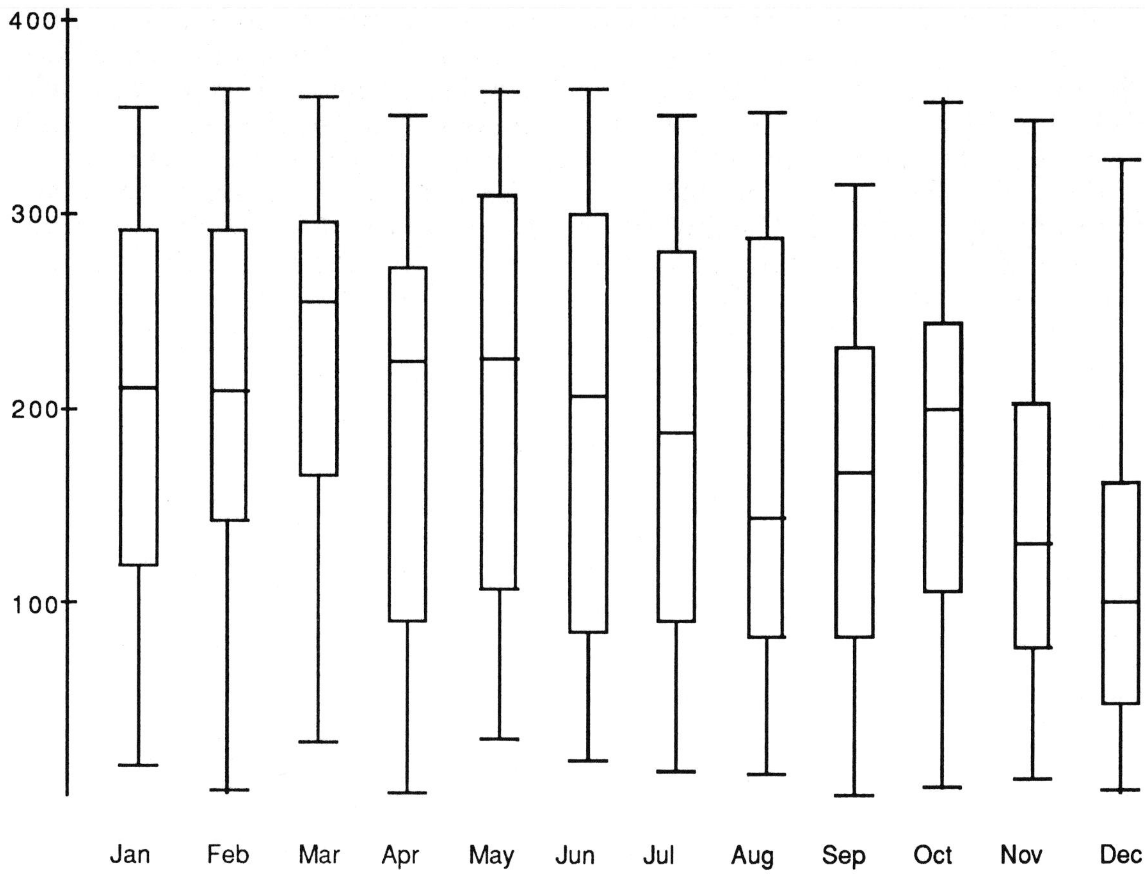
400
300
200
100
Jan
Feb
Mar
Apr
May
Jun
Jul
Aug
Sep
Oct
Nov
Dec

## Exercises

1. Why do we prefer to use the median rather than the mean draft number as the one-number summary for each month?

2. The side-by-side boxplots show the range in draft numbers for each month. Why is it better to use the IQR rather than the range as a summary of the variability within each month? If you did use the range instead of the IQR, what would you conclude?

## 2.2 Smoothing A Time Series

Some data are collected over time and come to us in the form of a *time series*. For example, you may record the high temperature in your home town each day for a year or the closing value of the Dow Jones Industrial Average each month for several years. The natural way to examine such a series of numbers is to construct a plot of the observations, in the vertical direction, against time, in the horizontal direction. Here is a plot of the number of household larcenies (in thousands) in the United States each month from January of 1973 through December of 1977.

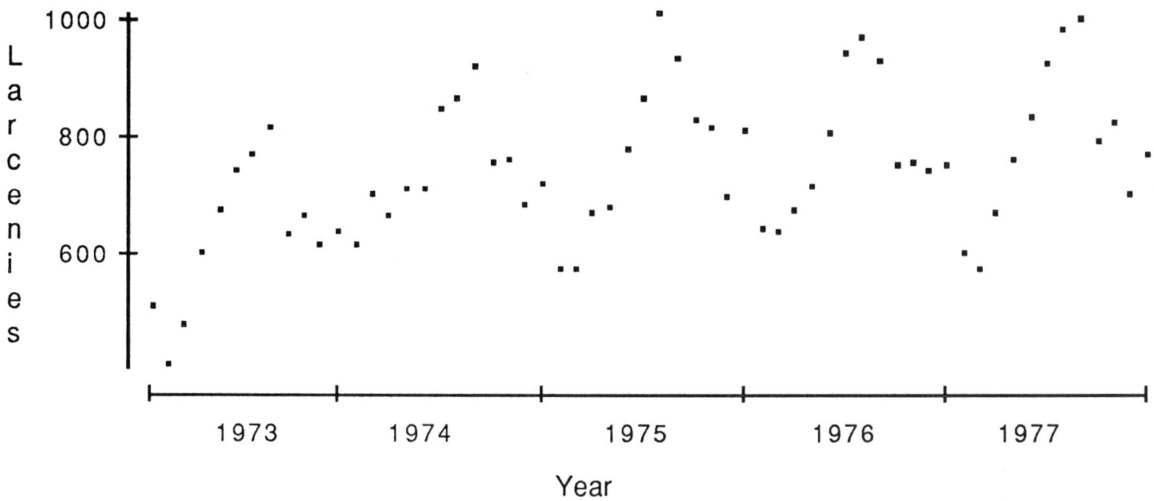


From this graph we can see a cyclical pattern in household larceny and an upward trend over time. The graph shows a strong seasonal pattern, with more larcenies being committed during summer months than during winter months. This may come as no surprise to you, since household larceny is theft of property or cash from a household, without the use of force, and summer months provide thieves with more opportunities than do winter months. Because seasonal effects in some data are very strong, statisticians often remove these effects and present "seasonally adjusted" data. For example, the Bureau of Labor Statistics makes seasonal adjustments to data such as the national unemployment rate.

The general theory of time series analysis is beyond the scope of this text, but we can consider some useful and easily applied methods for analyzing time series data. Many statistical computer packages include *smoothers*. A smoother is a statistical routine that eliminates the effects of extreme and unusual observations in a time series so that we can see underlying trends in the data. Here is a smoothed version of the larceny data.

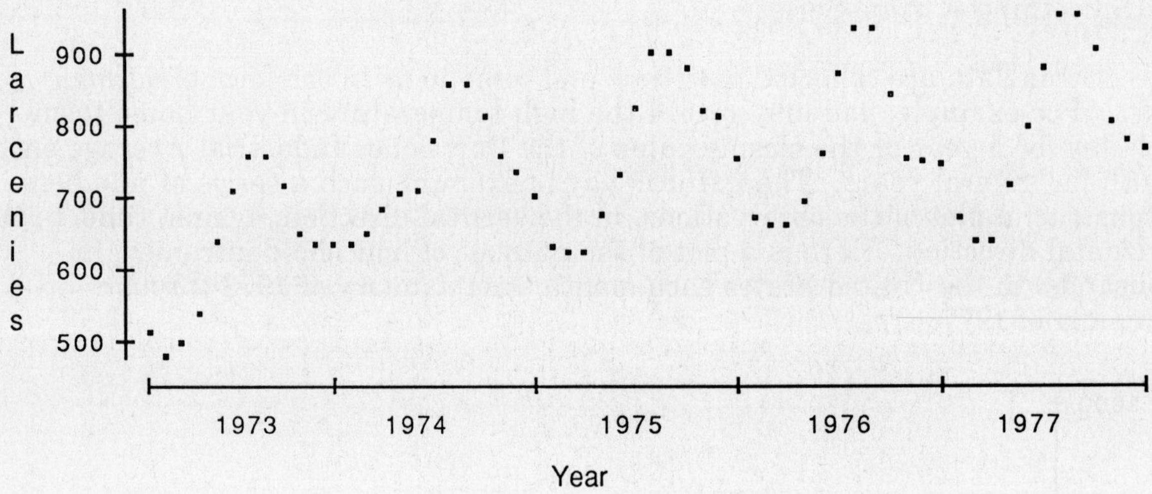


This smoothed plot shows the seasonal effects even more clearly than did the original plot.

As another example, consider the amount of water (in units of 100,000 gallons) that is processed each week by the water plant for the city of Oberlin, OH.

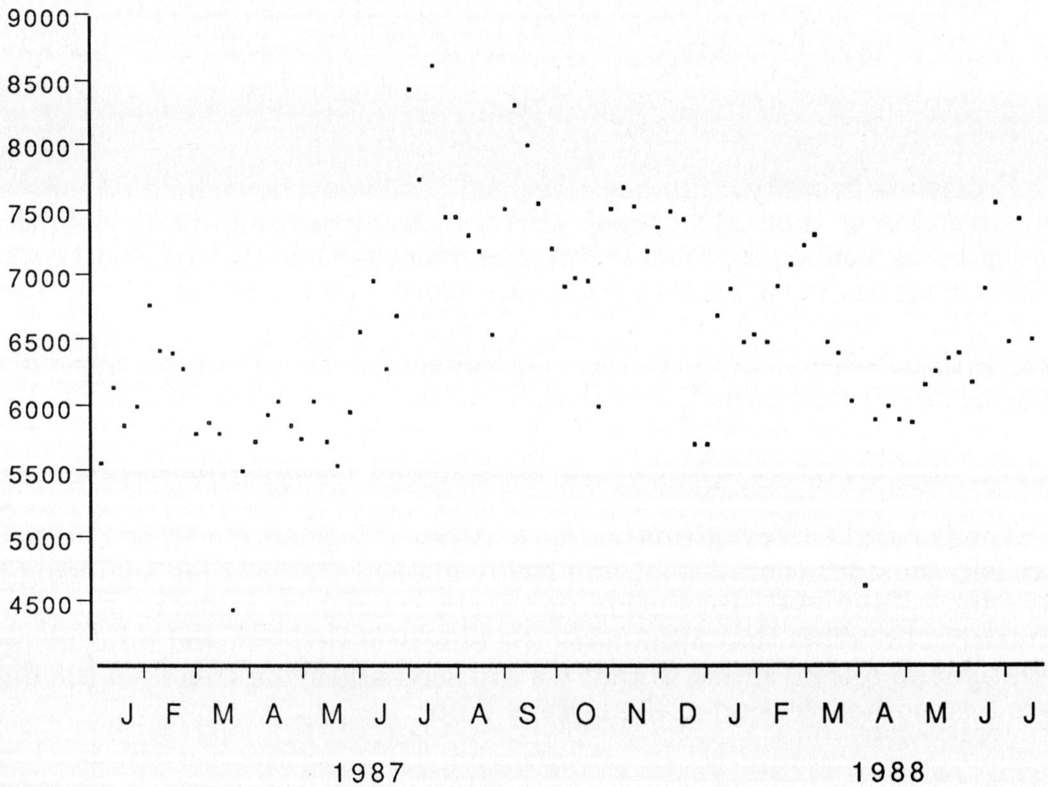


There are two extremely low points on this graph, which correspond to spring break at Oberlin College. Other relatively low points represent other weeks when the College is not in session (fall break, winter break). Smoothing the data eliminates these low points and helps us see a pattern.

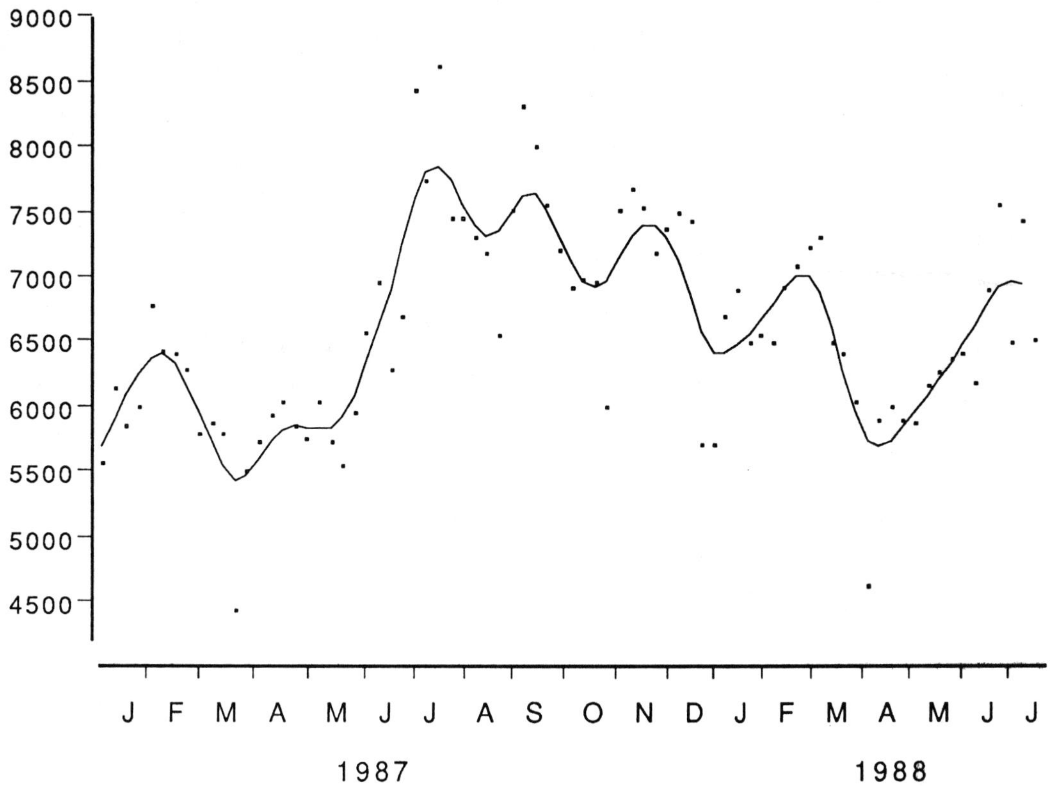


The following plot shows the average annual temperature in Cleveland for the years 1934-1987; the line running through the plot shows the data after *smoothing by running medians of 3.*

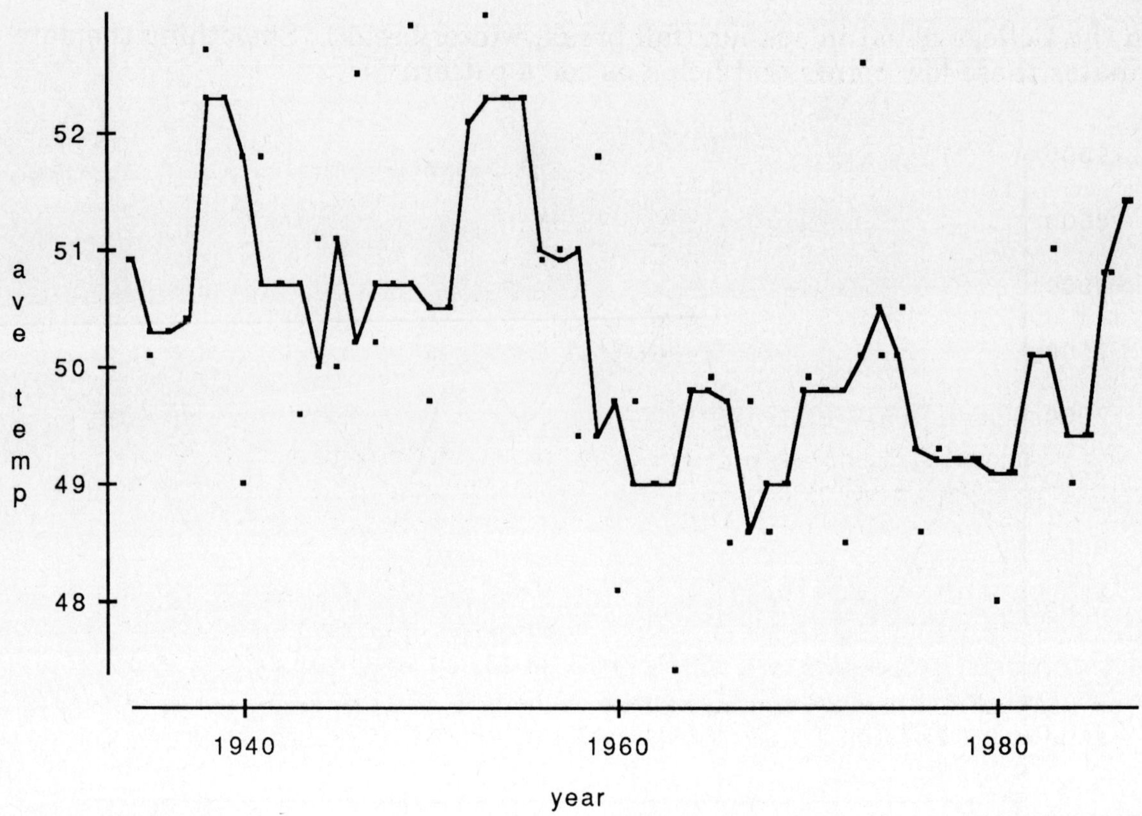


Smoothing involves replacing each observation in a time series by some combination of the observations that surround it. The simplest smoother consists of taking running medians of three observations at a time. This smoother replaces each observation by the median of a set of three numbers: the observation, the previous observation, and the next observation. Before applying running medians of three to the temperature data just shown, let's list the first few observations:

50.9 50.1 50.3 50.4 52.7 52.3 49.0 51.8

We begin by leaving the first observation as it is, since it is not in the middle of a set of three consecutive observations. (Likewise, we will not change the final observation. There are methods for dealing with the endpoints in a series, but they need not concern us.) We then imagine a window that is just wide enough to show us three observations at a time. We place the center of the window at the second observation and replace this observation the median of the three numbers in the window. Since the median of 50.9, 50.1, and 50.3 is 50.3, the second value in the smoothed series is 50.3:

| Original Data | Smoothed Data |
|---|---|
| 50.9 | 50.9 |
| 50.1 | 50.3 |
| 50.3 | |

Then we slide the window down one spot and replace the third observation by median of the three numbers now in the window:

| Original Data | | Smoothed Data |
|---|---|---|
| 50.9 | | 50.9 |
| 50.1 | ⎫ | 50.3 |
| 50.3 | ⎬ | 50.3 |
| 50.4 | ⎭ | |

We proceed in this fashion, "running" all of the way through the data. (When we get to the final observation we leave it unchanged.) This process produces the points shown by the solid line in the previous graph.

We could take the smoothed sequence of values and resmooth it by applying running medians of three a second time. We might repeat this process until resmoothing does not alter the data. This kind of repeated smoothing is fairly easy to do by hand, but we may prefer to use a computer and apply a *compound smoother* to a time series. This is discussed in the next section.

## 2.3 Complex Smoothers[1]

Smoothing a time series once, for example, by taking running medians of three, will remove many of the extraneous "bumps" from the underlying pattern. However, taking the smoothed series and applying another smoother to them will make further changes. Statisticians have found that *compound smoothers* - back-to-back applications of several different simple smoothers - can be quite effective. One such compound smoother[2] is called **4253H**, which stands for taking running medians of 4, then feeding the resulting series into a smoother of running medians of 2, followed by running medians of 5, then running medians of 3, then Hanning. Hanning consists of replacing each observation by a weighted average of the observation, the previous observation, and the following observation, using weights 1/4, 1/2, and 1/4. Here is a step-by-step application of 4253H to the first part of the larceny data.

We begin by smoothing the original data with running medians of four. We leave the first point unchanged and replace the second point with the average of the first two points. (We handle the last two points in the same way.) Then we move through the data with a window that allows us to see four observations at a time and we take the median of each set four consecutive points (the average of the two points that are in the middle when the four are sorted from smallest to largest). This yields the column labeled **4** in Table 1. For example, the value 495 in column **4** is the median of 511, 408, 479, 598, since (479 + 511)/2 = 495.

The next step is to apply running medians of two to the sequence just produced. We leave the first observation unchanged and, starting with the first pair of points, take medians of pairs of points. (The median of two points is their average.) This produces the column labeled **2** in Table 1.

Now we apply running medians of five to this series. We leave the first value unchanged and replace the second value by the median of the first three values. We replace the third value by the median of the first five values. We continue to use a window of size 5 until we near the last few points, which we handle in the same way that we handled the first few points. This yields column **5**. Then we create column **3** by applying running medians of three to the data in column **5**. The final part of the process is Hanning. We replace each point, $Y_t$, in column **3** by

$$.25Y_{t-1}+.5Y_t+.25Y_{t+1}.$$

This yields column **H**.

---

[1] The material here on complex smoothers is, indeed, a bit complex. Within the statistics community there is less widespread agreement about how and when to use these tools than is the case for the other ideas presented in this book.

[2] This smoother has been promoted by John Tukey, who originally developed many of the techniques of exploratory data analysis.

Table 1

| Original Data | 4 | 2 | 5 | 3 | H |
|---|---|---|---|---|---|
| | 511 | | | | |
| 511 | | 511 | 511 | 511 | 511 |
| | 459.5 | | | | |
| 408 | | 477.25 | 511 | 511 | 512.4375 |
| | 495 | | | | |
| 479 | | 516.75 | 516.75 | 516.75 | 532.9375 |
| | 538.5 | | | | |
| 598 | | 587.25 | 587.25 | 587.25 | 590.6875 |
| | 636 | | | | |
| 674 | | 671.5 | 671.5 | 671.5 | 665.25 |
| | 707 | | | | |
| 740 | | 730.75 | 730.75 | 730.75 | 715.9375 |
| | 754.5 | | | | |
| 769 | | 754.5 | 730.75 | 730.75 | 730.75 |
| | 754.5 | | | | |
| 812 | | 735.5 | 730.75 | 730.75 | 718.75 |
| | 716.5 | | | | |
| 634 | | 682.75 | 682.75 | 682.75 | 684.5 |
| | 649 | | | | |
| 664 | | 641.75 | 641.75 | 641.75 | 651.625 |
| | 634.5 | | | | |
| 614 | | 629.5 | 640.25 | 640.25 | 640.625 |
| | 624.5 | | | | |
| 635 | | 624.5 | 640.25 | 640.25 | 640.25 |
| | 624.5 | | | | |
| 614 | | 640.25 | 640.25 | 640.25 | 654.6875 |
| | 656 | | | | |
| 698 | | 698 | 698 | 698 | 698 |
| | 698 | | | | |

Some statistical software packages include smoothers. For example, the compound smoother *4253H,twice* is available in MINITAB. (This smoother is a bit more complex than the 4253H smoother just described. If we take the original data and subtract the results of the 4253H smoother (e.g., the "original data" column minus the **H** column in Table 1) we get a column of *residuals*. Twicing involves applying the 4253H smoother to the residuals, to get smoothed residuals, and then adding the smoothed residuals to the smoothed data.)

Another common method for examining a time series is to use an *exponential smoother*. Here we replace each observation, $Y_t$, by a weighted average of the preceding observations, giving exponentially less and less weight to observations that are further and further away from the current observation. Specifically, we create a new series by setting

$$F_{t+1} = \lambda Y_t + (1-\lambda)F_t,$$

where l is a fixed parameter between 0 and 1. We start with $F_2 = Y_1$. Then

$$F_3 = \lambda Y_2 + (1-\lambda)F_2 = \lambda Y_2 + (1-\lambda)Y_1$$

and

$$F_4 = \lambda Y_3 + (1-\lambda)F_3 = \lambda Y_3 + (1-\lambda)[lY_2 + (1-\lambda)F_2]$$
$$= \lambda Y_3 + (1-\lambda)\lambda Y_2 + (1-\lambda)^2 Y_1.$$

Likewise,

$$F_5 = \lambda Y_4 + (1-\lambda)\lambda Y_3 + (1-\lambda)^2\lambda Y_2 + (1-\lambda)^3 Y_1.$$

The procedure is called exponential smoothing because the weights $\lambda$, $(1-\lambda)\lambda$, $(1-\lambda)^2\lambda$,... decrease exponentially.

## 3.1 Normal Probability Plots

Many statistical techniques are based on the assumption that the data at hand come from a normal distribution. There are several ways to check this assumption with any given set of data: we could construct a histogram or stem-leaf diagram of the data, we could generate a dotplot, or we could generate a *normal probability plot.*

**EXAMPLE 1**

Many people exercise by swimming. A lifeguard at an indoor pool observed several swimmers and recorded the amount of time that each spent in the water. Here are the times, in minutes, for a sample of 15 swimmers[1]:

19, 18, 38, 39, 24, 25, 29, 22, 28, 43, 34, 10, 32, 5, 30.

The average time is 26.4 minutes, but there is considerable variability in the data. A stem-leaf diagram shows that the distribution is nearly bell-shaped

```
  1   0 | 5
  4   1 | 089
(5)   2 | 24589
  6   3 | 02489
  1   4 | 3
```

We might infer that swimming times follow a normal curve. The stem-leaf diagram supports this inference, since we can see a bell-shape in the distribution. However, there is a better visual tool available to help us assess normality. Here is a normal probability plot of the same data:

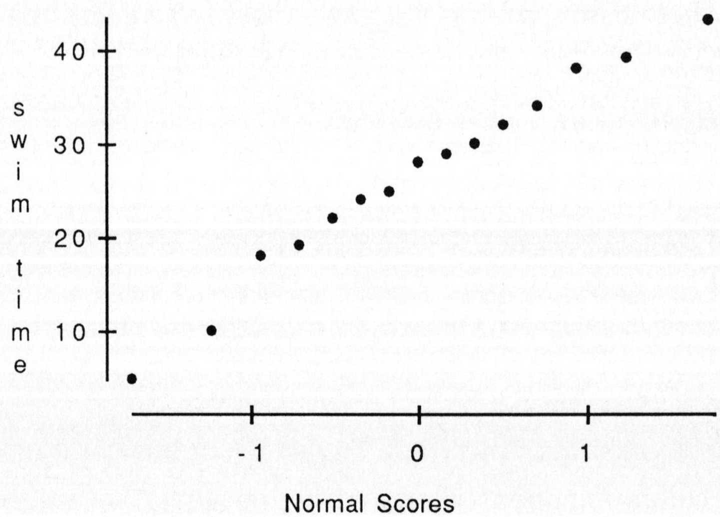


[1] These data were collected by Brian Patterson at Carr Pool in Oberlin, OH, on the evening of 13 October 1987.

This plot supports the assumption that swim times are normally distributed because the points fall (more or less) on a straight line.

Judging whether or not points fall on a straight line is easier than judging whether or not a distribution is bell-shaped; thus, a normal probability plot is the preferred tool for assessing normality. Many statistical computer packages can be used to produce normal probability plots, which are also known as normal quantile plots or Q-Q plots.

Normal probability plots are very difficult to construct by hand, and there are several variations on the idea of a normal probability plot. Thus, it is more important to learn how to interpret such plots than it is to learn precisely how they are constructed. Nonetheless, here is the general idea behind generating such a plot: Suppose we took a random sample of 15 observations from a standard normal distribution (i.e., with average zero and standard deviation one). Clearly, we would expect the sample mean and median to be somewhere around zero and the sample standard deviation to be near one. We would expect the largest observation to be somewhat greater than one.

We can ask "How large should we expect the largest observation to be (in a sample of size 15 from a standard normal distribution)? How large should the second-largest observation be? etc." These theoretical values are called *normal scores*. Here is a list of the observed swimming times, ranked from smallest to largest, together with the corresponding normal scores.

| rank | observed value | normal score |
|---|---|---|
| 1 | 5 | -1.71 |
| 2 | 10 | -1.23 |
| 3 | 18 | -0.94 |
| 4 | 19 | -0.71 |
| 5 | 22 | -0.51 |
| 6 | 24 | -0.33 |
| 7 | 25 | -0.16 |
| 8 | 28 | 0.00 |
| 9 | 29 | 0.16 |
| 10 | 30 | 0.33 |
| 11 | 32 | 0.51 |
| 12 | 34 | 0.71 |
| 13 | 38 | 0.94 |
| 14 | 39 | 1.23 |
| 15 | 43 | 1.71 |

Notice that the normal scores are symmetric about zero, which is the 8th largest expected observation.

Now consider the plot of the observed values vs. the normal scores. If the data really are normally distributed, then we expect the gaps between observed values, Y, to correspond to the theoretical gaps between the normal scores, X. This will result in a plot that looks like a straight line.

We do not expect the normal probability plot to be perfectly linear, but marked departures from linearity suggest that the data are not from a normal distribution. Judging normal probability plots takes some experience. Here are normal probability plots for six samples of data that really are from a normal distribution. Note that none of these plots is perfectly linear.

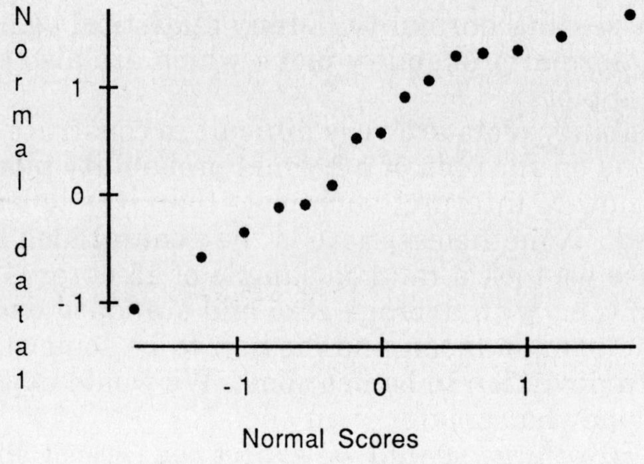


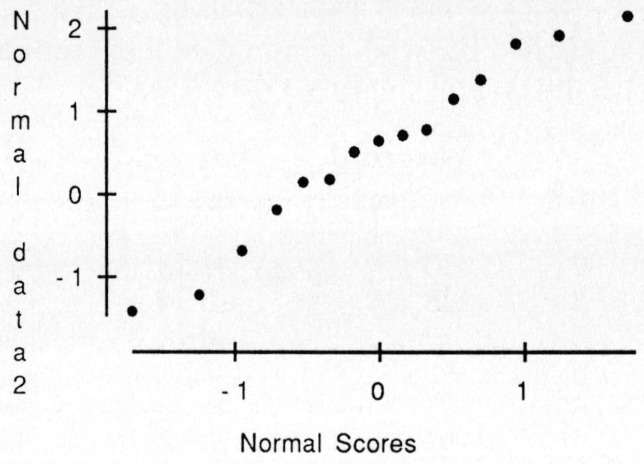


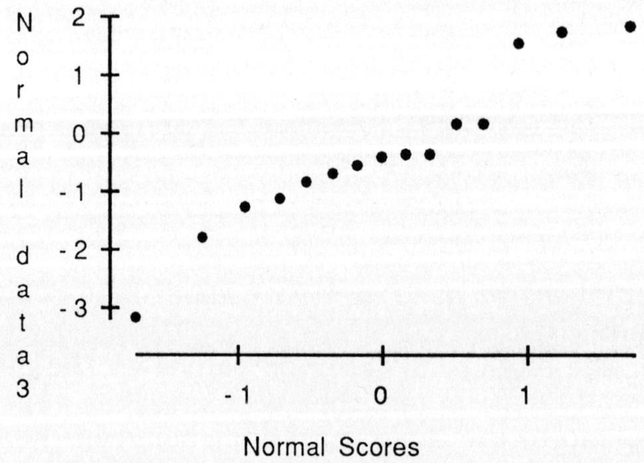

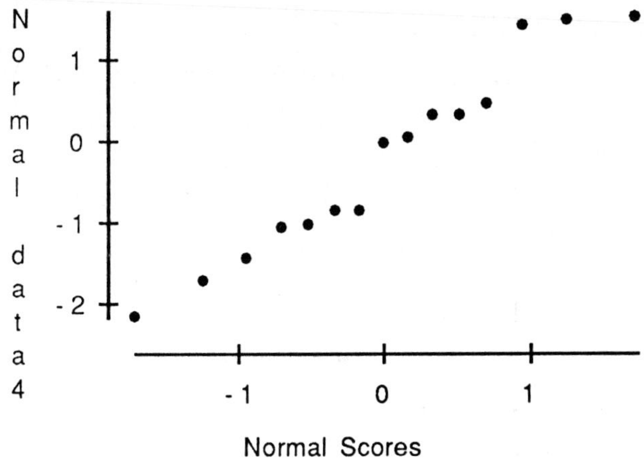
Normal data4
1
0
-1
-2
-1
0
1
Normal Scores

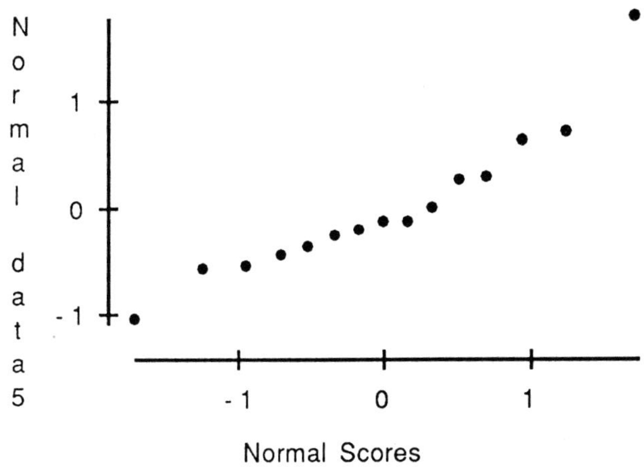
Normal data5
1
0
-1
-1
0
1
Normal Scores

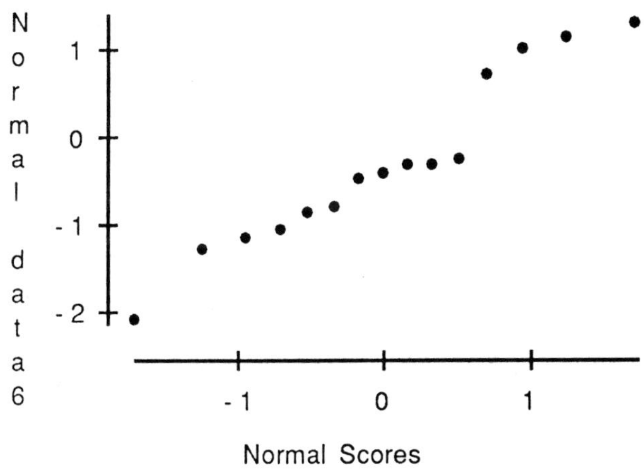
Normal data6
1
0
-1
-2
-1
0
1
Normal Scores

**EXAMPLE 2**

As an example of truly non-normal data, here is a histogram of the winning margin (winner's score - loser's score) for NFL football games played in 1987.

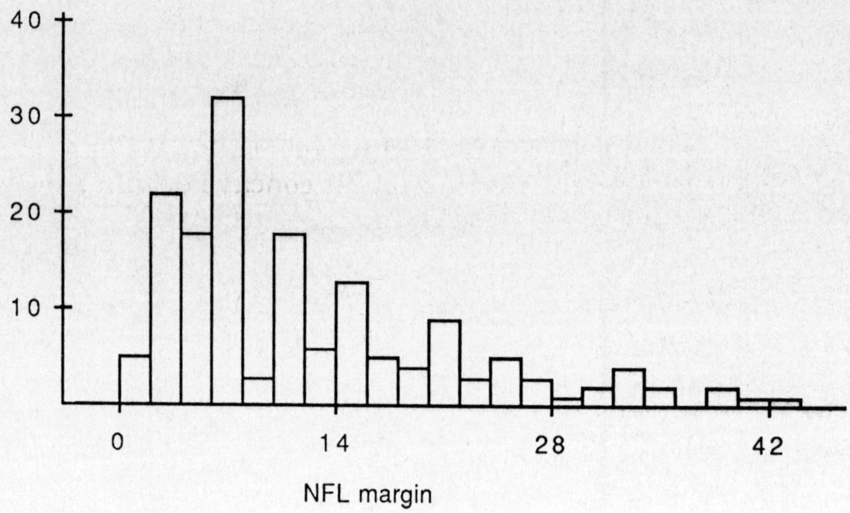


The normal probability plot shows that this distribution is not bell-shaped. The long right-hand tail in the distribution causes the plot to flatten on the left.

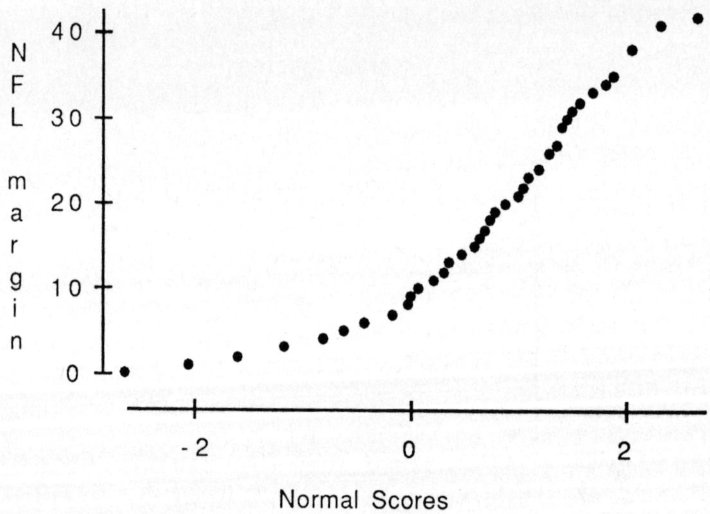

**EXAMPLE 3**

Here is a dotplot of the number of home runs that Henry Aaron hit each in each year of his career.

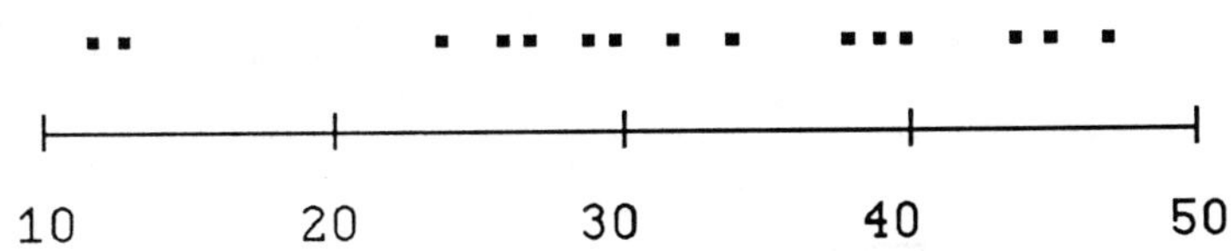


This distribution is slightly skewed to the left (has a long left-hand tail), causing the normal probability plot to be a bit concave down. However, this plot is nearly as linear as are some of the plots on pages 34 and 35.

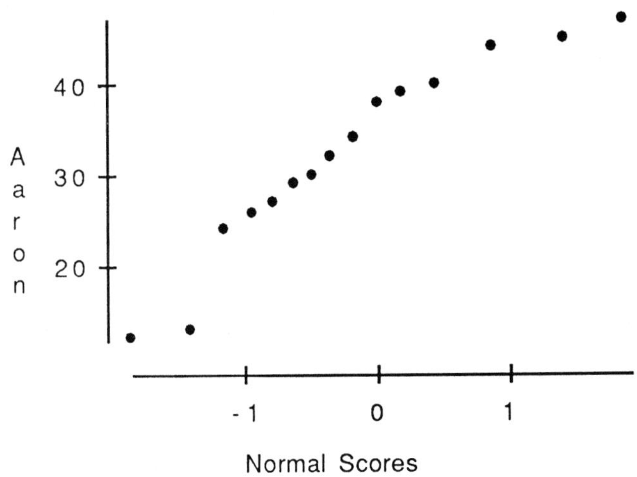


<u>Exercise</u>

1. Here are June precipitation totals for the city of Bellefontaine, Ohio, for the years 1978-1989:

| Rain | Normal scores |
|---|---|
| 7.05 | 1.10 |
| 5.43 | -1.10 |
| 1.94 | 0.53 |
| 4.14 | -0.10 |
| 8.44 | -0.79 |
| 2.17 | -1.61 |
| 1.70 | 0.31 |
| 3.23 | 0.10 |
| 1.24 | 1.61 |
| 3.94 | 0.79 |
| 5.75 | -0.53 |
| 2.29 | -0.31 |

Construct a histogram and a normal probability plot for these data. Comment on the shape of the distribution.

3.2 Transformations

We saw in section 3.1 how to make a visual check of the assumption that a set of data come from a normal distribution. Sometimes we wish to use a statistical technique that depends on the assumption that the data are normally distributed. We can often transform non-normal data to make them approximately normal. Computers make it easy to transform data and to check, with normal probability plots, whether the transformed data satisfy the assumption of normality.

**Example 1**

Roberta Foster often finds coins on the ground as she strolls around Springfield, Ohio. The follow data are the numbers of coins she has found in each of ten years:

62, 35, 23, 31, 24, 37, 29, 39, 50, 44.

The dotplot and normal probability plot show that this distribution is skewed - the value 62 is quite a bit larger than any other value. Notice that skewed nature of the data is reflected in the normal probability plot, which is curved upward .

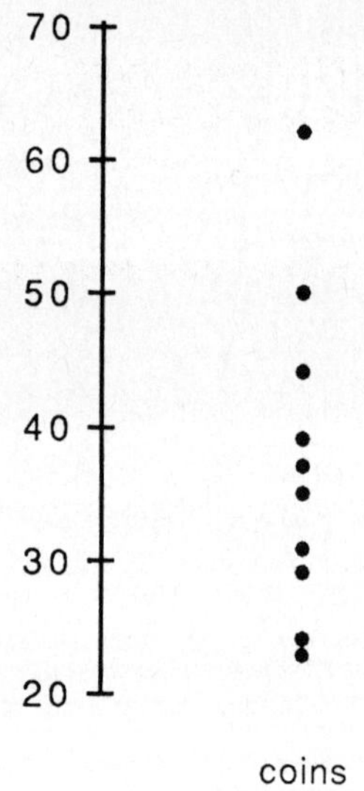


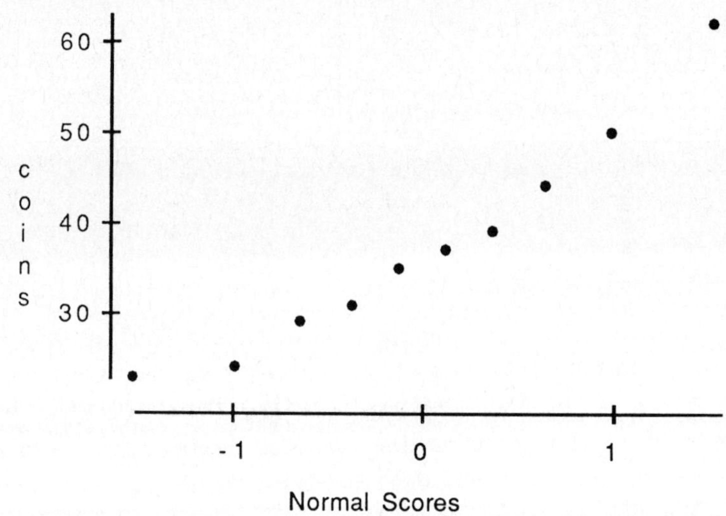


To make this distribution symmetric we need to "pull in" the upper tail. One attempt is to take the square root of each value. This changes the original data, X, to the transformed data, Y:

7.87, 5.92, 4.80, 5.57, 4.90, 6.08, 5.39, 6.24, 7.07, 6.63.

The new dotplot and normal probability plot show that the distribution is closer to being normal than was the original distribution.

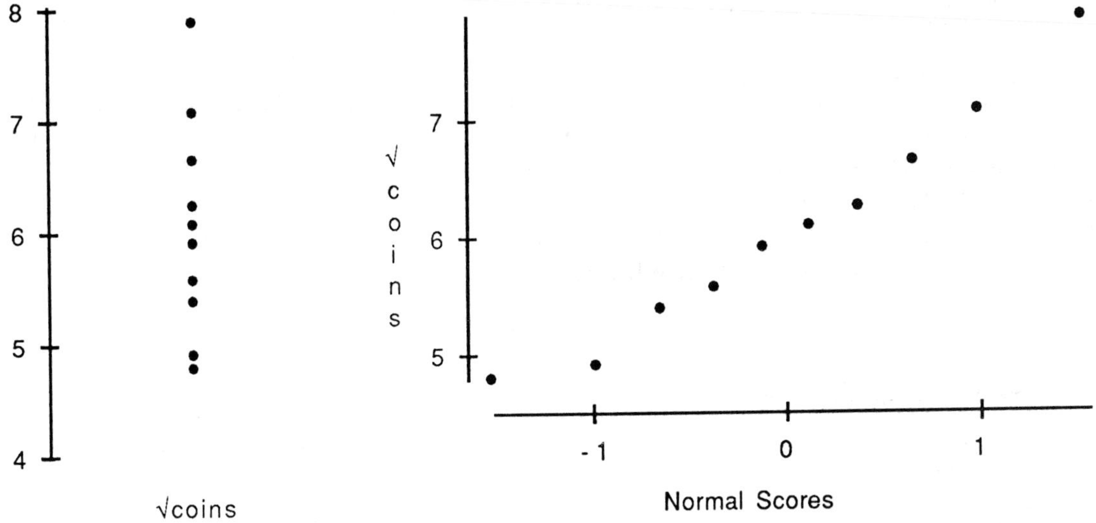


The square root transformation is an example of a *power transformation*. To use a power transformation, we choose a power, such as .5, and raise each value to that power:

$$Y = \sqrt{X} = X^{.5}.$$

Using a power less than 1 pulls in the upper tail, while pushing out the lower tail; powers greater than 1 do the opposite. Since the coin data have a long upper tail (the normal probability plot is concave up), we choose from among powers less than 1.

The following plots show the result of taking the logarithm of each value along with the results of using the powers -.5, and -1.

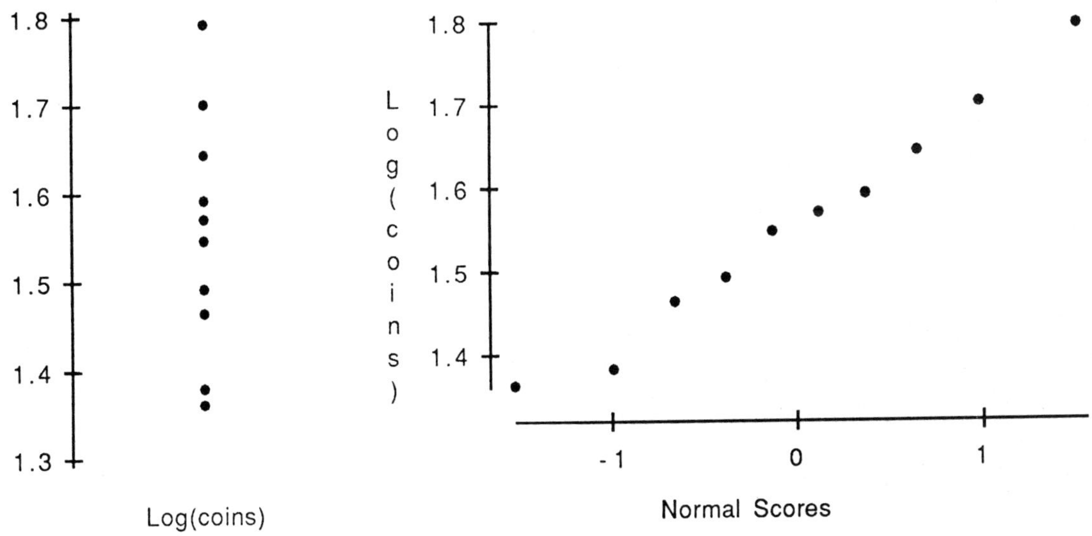

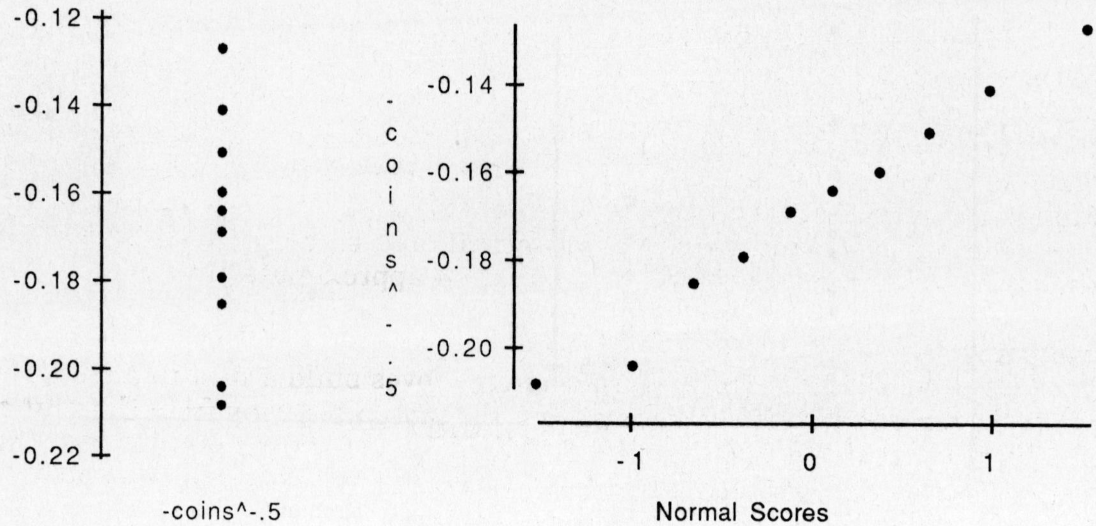


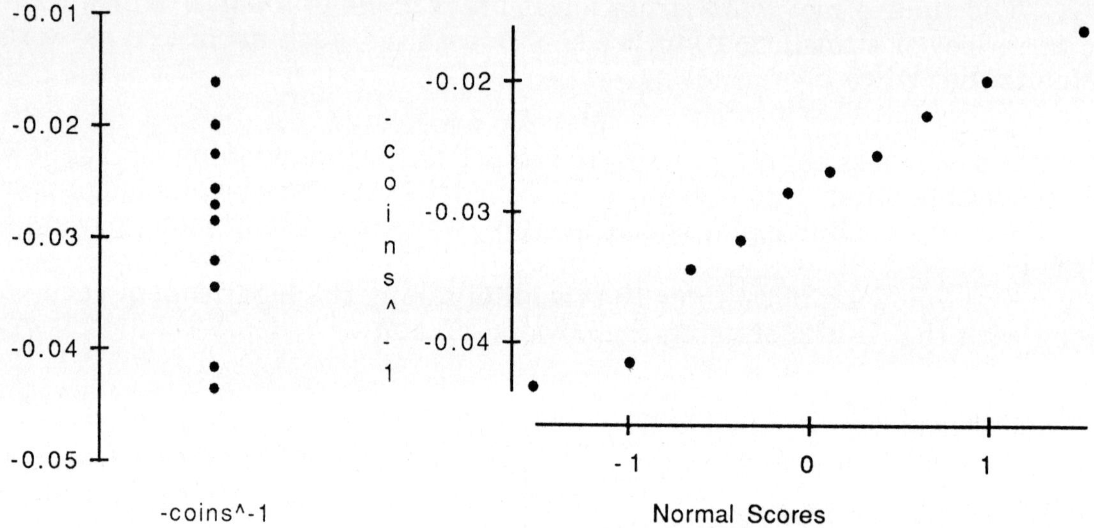


The logarithm transformation is widely used. If we use a power transformation with a power close to 0 the result is close to what we get by taking logs[1]. Thus, we can think of the logarithm transformation as corresponding to the value 0 in a decreasing series of powers (.5, .25, log, -.25, -.5, -.75, etc.)[2]. (Note: We can use any power we like, such as .2387. However, it is common to consider powers that are "round numbers," such as .5, and .25.) I used log base 10, but using a different base, such as e=2.7182... (i.e., taking natural logs), has the same effect. That is, when taking logs switching from one base to another only changes the scale, but not the shape, of the distribution.

---

[1] If we define $Y_p(X) = (X^p - 1)/p$, then as p approaches zero, $Y_p(X)$ approaches the natural logarithm of X. The family of transformations $Y_p(X)$ was proposed by George Box and David Cox in "An analysis of transformations (with discussion)," *J. R. Statist. Soc. Ser. B*, **26** (1964), 211-246.

[2] Note that we can not use a power less than or equal to zero if we have count data that take on the value of zero (e.g., we can not take the logarithm of zero). In this case, we can add a small constant, such as .5, to each observation before applying the transformation.

When using negative powers, it is convenient to multiply the results by -1 to preserve the order of the data. For example, when I used the power -1 above, the largest value of X (62) became the smallest value of Y (1/62). Thus, I multiplied the transformed values by -1 to put them back into their original order, while not changing the shape of the distribution. This makes it easier to compare the dotplot and normal probability plot labeled "-coins^-1" to the preceding plots.

The reciprocal square root transformation (i.e., using the -.5 power) does a fairly good job of making the coin distribution approximately normal.

**Example 2**

One day I watched a pair of mourning doves build a nest in a tree near my house. I recorded the length of time that the male was away from the nest on 73 trips he made to bring twigs, straw, and the like back to the female (who actually constructed the nest). Many of these trips were quite short, but some were rather long. The following histogram displays the data.

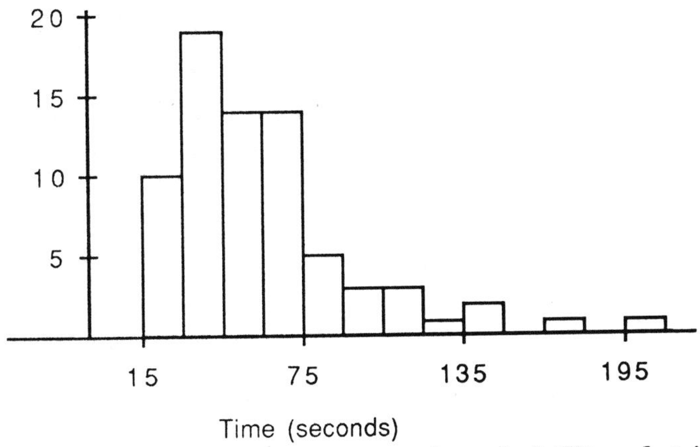


Note the long upper tail here. The normal probability plot is clearly not linear. Rather, it curves upward, reflecting the fact that the histogram has a long upper tail.

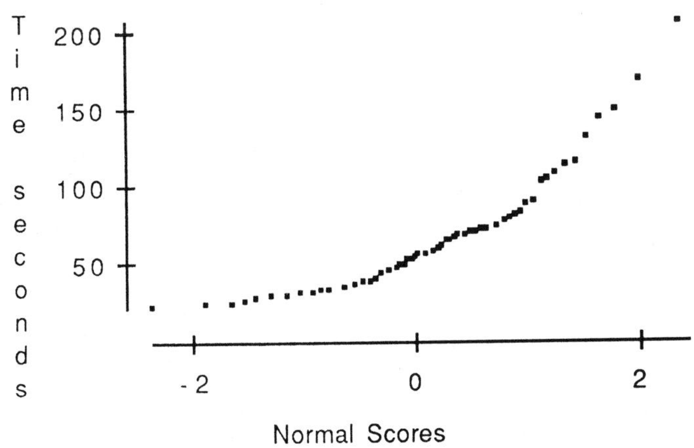

To make the distribution bell-shaped we must use a transformation that pulls in the upper tail. Taking logs is an improvement, but leaves the distribution slightly skewed to the right.

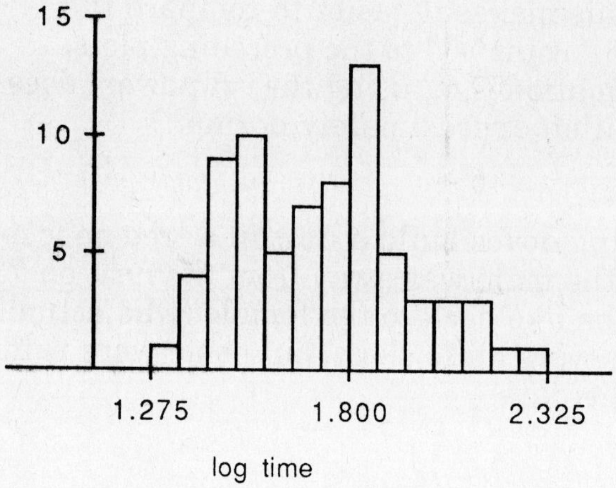


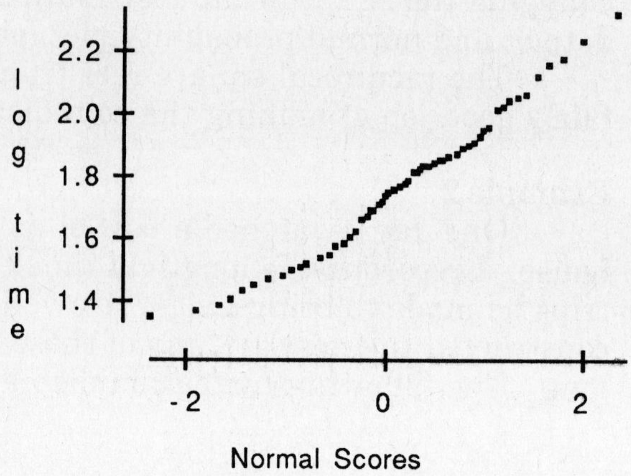


A stronger transformation pulls in the upper tail even more. Using the power -.5 (i.e., taking the reciprocal of the square root of each value) gives us a more nearly symmetric distribution.

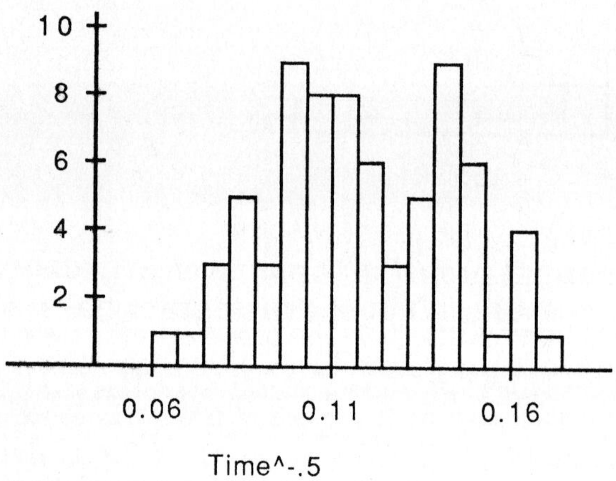


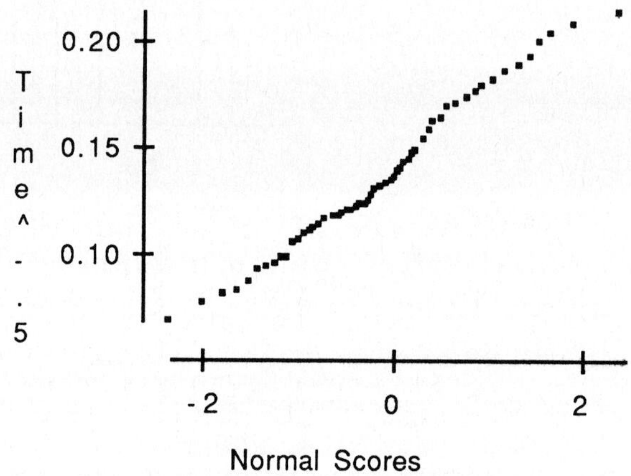


**Example 3**

Linguists are sometimes interested in the distribution of the number of words per sentence of an author's work. In 1940 C. B. Williams examined *The Work, Wealth and Happiness of Mankind*, by H. G. Wells[3]. He took a sample of 600 sentences and found that the average number of words per sentence was 24.1. (This is quite a bit less than the average of 31.2 that he found for *An Intelligent Woman's Guide to Socialism*, by G. B. Shaw.)

The histogram and normal probability plot below show that this distribution is skewed to the right, as there were some sentences that were much longer than

[3] C. B. Williams (1940) "A note on the Statistical Analysis of Sentence Length as a Criterion of Literary Style," *Biometrika* 31:356-361.

average. (The shortest sentence in the sample had 2 words, while there were 13 sentences with more than 60 words.)

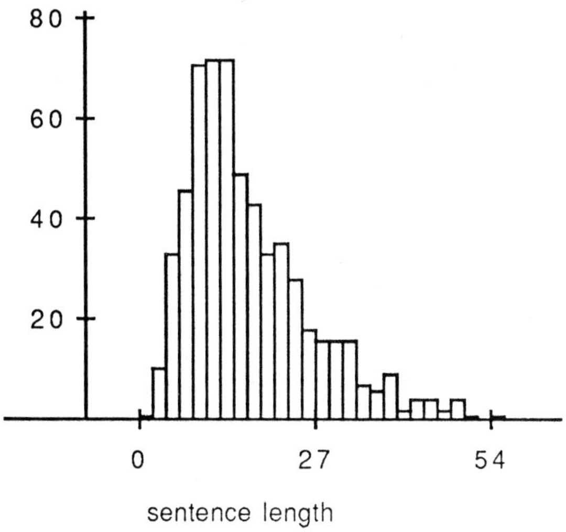


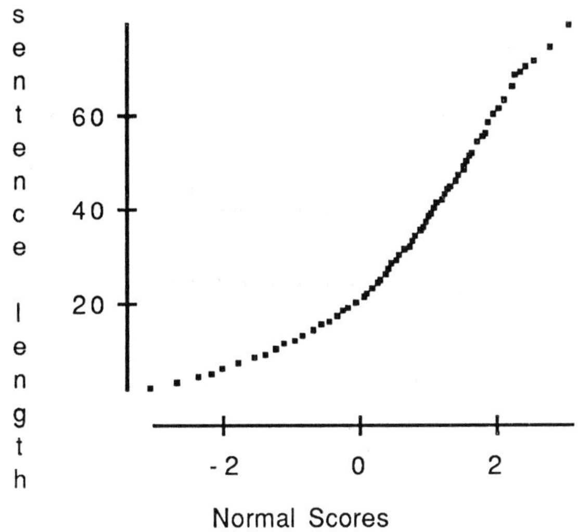


Taking the cube root of each sentence length (i.e., raising each value to the power .33) produces a distribution that is quite close to normal.

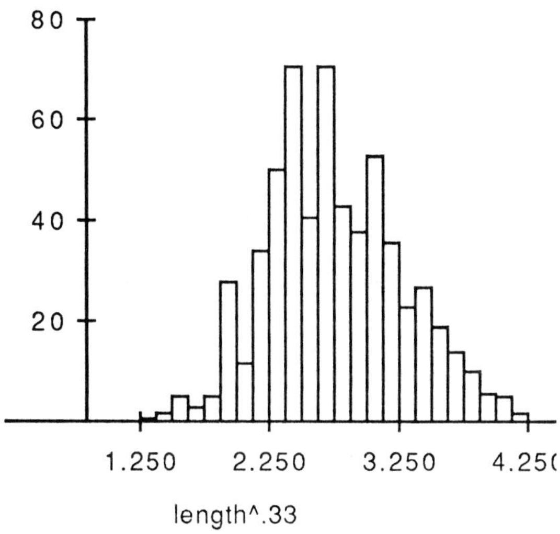


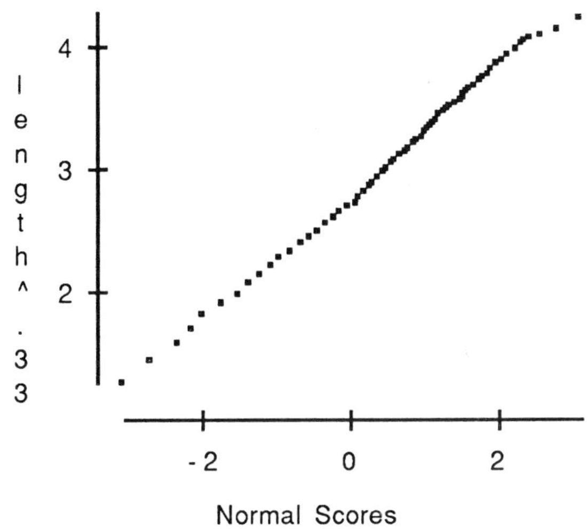

We can compare Wells to Shaw by looking at side-by-side boxplots of the original (untransformed) data:

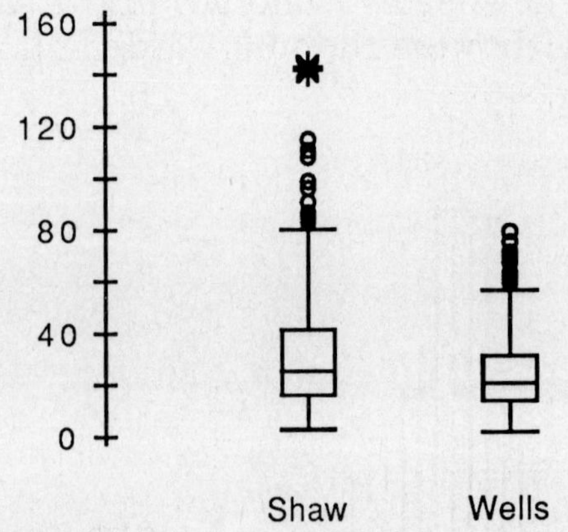


However, it is a bit easier to see how the authors compare when we examine boxplots of the transformed data:

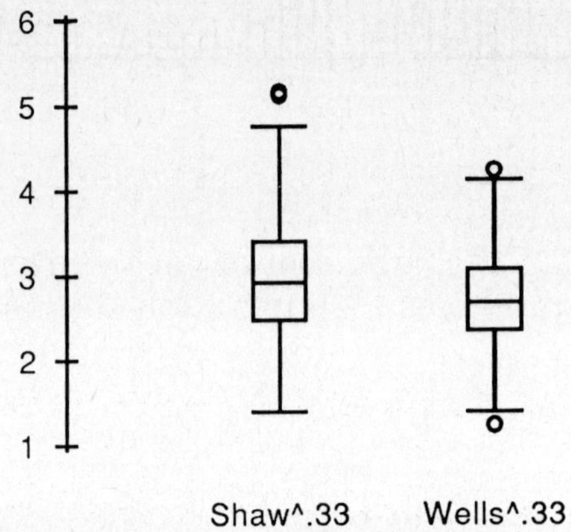


**Example 4**

As an example of a distribution that is skewed to the left, consider the number of credits that students earn in the first year of college. A histogram and normal probability plot are given below for a sample of 50 students from Oberlin College. The normal probability plot is concave down, reflecting the fact that the histogram has a long lower tail.

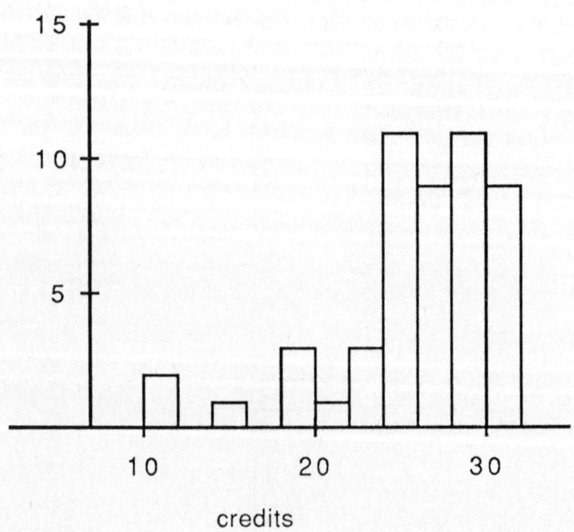


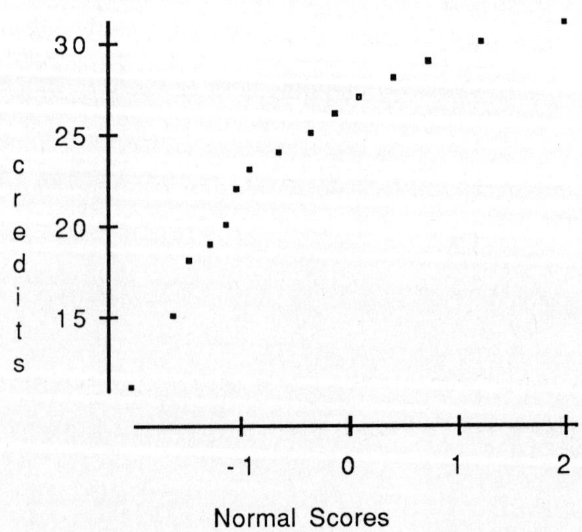

To pull in the lower tail, we raise each observation to a power greater than 1. Raising each value to the fourth power makes the distribution approximately normal, as the following plots show.

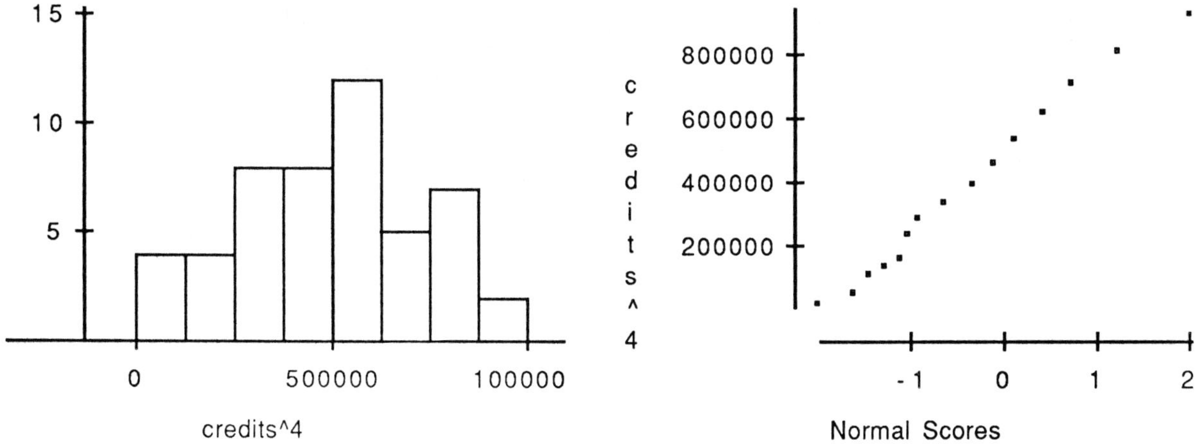


Exercises

1. Consider the precipitation data from exercise 1 on page 37. Transform these data to get a symmetric distribution.

2. Planes land at O'Hare airport in Chicago at a fairly steady rate. There is, however, some variability in the times between arrivals of planes on any given runway. Here are interarrival times, in seconds, on one runway for a sample of size 15 taken during the late afternoon on 3 March 1990:

102, 116, 47, 82, 74, 68, 48, 79,
137, 164, 112, 57, 76, 91, 141.

A histogram and normal probability plot of these data are given below and show that this distribution is skewed to the right. The average time between planes in this sample is 92.9 seconds, while the median is 82 seconds; the standard deviation is 35.1 seconds.

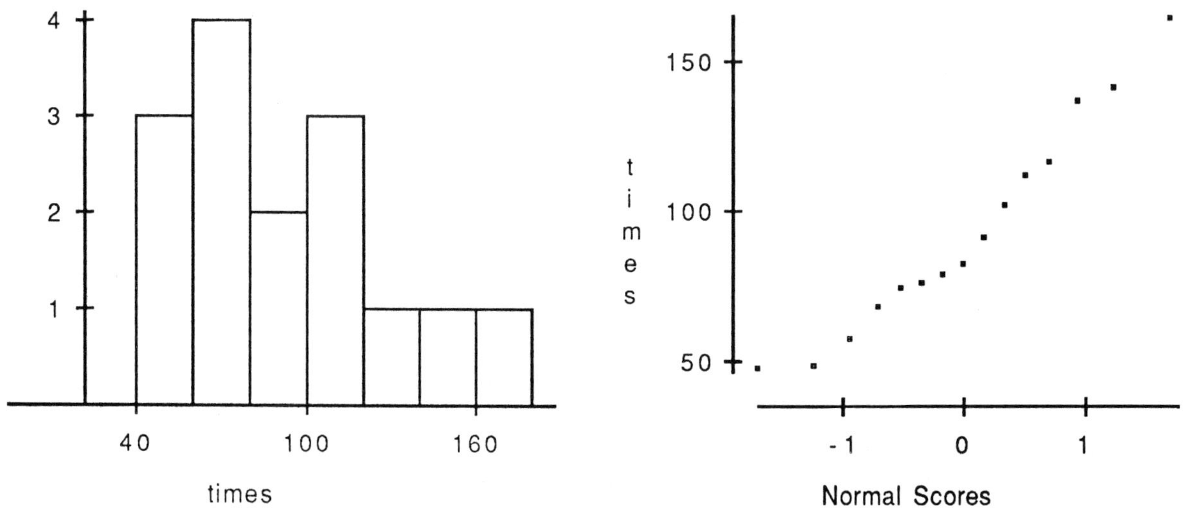


Transform these data to get a symmetric distribution.

## 4.1 Scatter Plots

The most powerful tool that statisticians use when studying the relationship between two variables is the *scatter plot*. A scatter plot can be worth a thousand words, and in the case of the Space Shuttle Challenger disaster, much more than that.

**Example 1** *The Challenger Accident*.
On 28 January 1986, the twenty-fifth flight of the U.S. Space Shuttle program ended in disaster when one of the rocket boosters of the Shuttle *Challenger* exploded shortly after lift-off, killing all seven crew members. The Presidential Commission that investigated the accident concluded that it was caused by the failure of an O-ring in a field joint on the rocket booster, and that this failure was due to a faulty design that made the O-ring unacceptably sensitive to a number of factors, including temperature[1]. The Commission learned that O-rings had been damaged on many of the twenty-four previous flights. Here is a scatter plot of two variables, X=joint temperature (in degrees Fahrenheit) and Y=number of incidents of "thermal distress" to field joint O-rings, measured on Shuttle flights prior to the accident. (The data are missing from the one flight - the hardware from the fourth flight was lost at sea - so only 23 flights are represented in the plot. There were incidents of damage to O-rings that were not on field joints, but only field joint incidents are included in the following plot.)

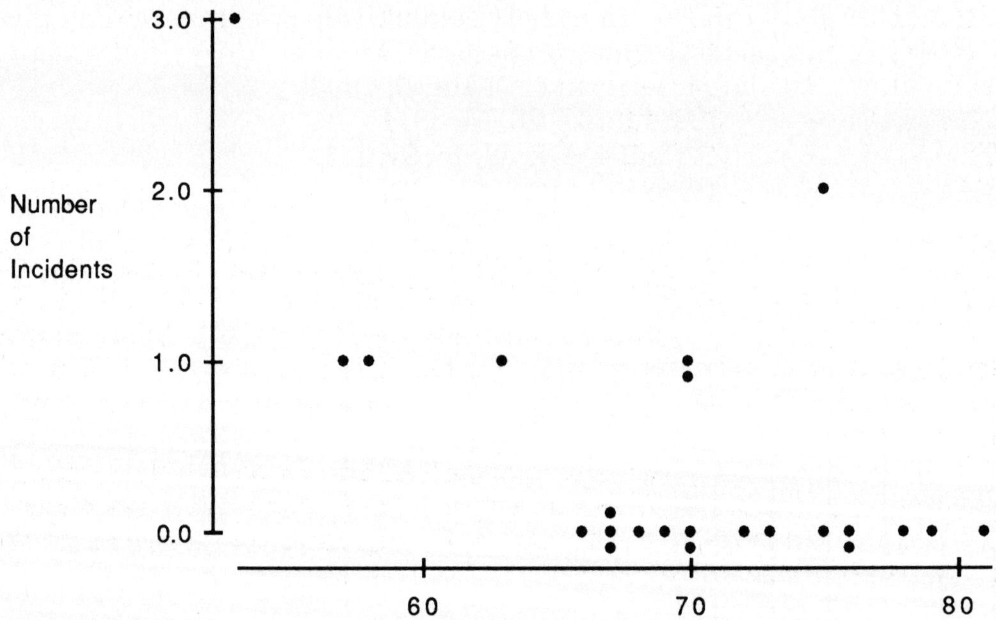


[1]*Report of the Presidential Commission on the Space Shuttle Challenger Accident* (1986).

This scatter plot is a visual summary of the following collection of numbers.

| Flight | Date | Temperature | Number of Damage Incidents |
|---|---|---|---|
| STS-1 | 4/12/81 | 66 | 0 |
| STS-2 | 11/12/81 | 70 | 1 |
| STS-3 | 3/22/82 | 69 | 0 |
| STS-4 | 6/27/82 | 80 | (data not available) |
| STS-5 | 1/11/82 | 68 | 0 |
| STS-6 | 4/4/83 | 67 | 0 |
| STS-7 | 6/18/83 | 72 | 0 |
| STS-8 | 8/30/83 | 73 | 0 |
| STS-9 | 11/28/83 | 70 | 0 |
| STS 41-B | 2/3/84 | 57 | 1 |
| STS 41-C | 4/6/84 | 63 | 1 |
| STS 41-D | 8/30/84 | 70 | 1 |
| STS 41-G | 10/5/84 | 78 | 0 |
| STS 51-A | 11/8/84 | 67 | 0 |
| STS 51-C | 1/24/85 | 53 | 3 |
| STS 51-D | 4/12/85 | 67 | 0 |
| STS 51-B | 4/29/85 | 75 | 0 |
| STS 51-G | 6/17/85 | 70 | 0 |
| STS 51-F | 7/29/85 | 81 | 0 |
| STS 51-I | 8/27/85 | 76 | 0 |
| STS 51-J | 10/3/85 | 79 | 0 |
| STS 61-A | 10/30/85 | 75 | 2 |
| STS 61-B | 11/26/85 | 76 | 0 |
| STS 61-C | 1/12/86 | 58 | 1 |
| STS 51-L | 1/28/86 | 31 | (Challenger accident) |

Each flight is represented by a point whose X-axis coordinate is Temperature and whose Y-axis coordinate is Number of Damage Incidents. Notice that as temperature (X) goes down, number of incidents (Y) tends to go up, suggesting that low temperatures cause problems for the O-rings. Indeed, this relationship is rather striking. All of the flights launched below 65 degrees had at least one damage incident.

Unfortunately, when managers discussed the possible effects of cold weather, on the night before the *Challenger* launch, they considered only the seven flights for which thermal distress of O-rings had been observed. The scatter plot for these flights does not show an obvious trend.

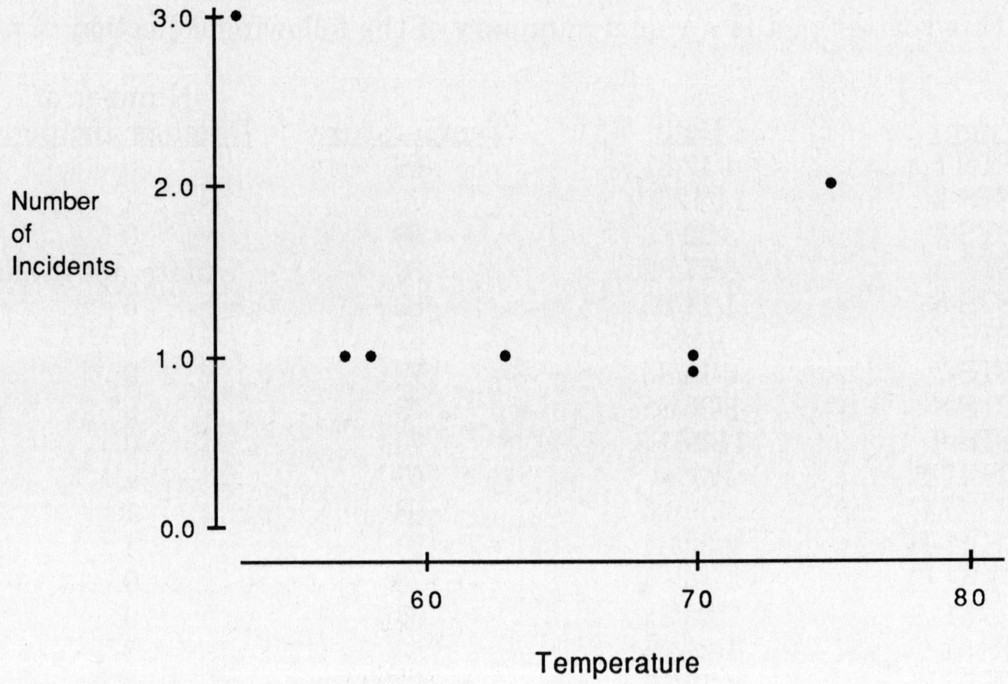


Compare this plot to the first plot, which shows all of the data. The scatter plot of the complete set of data is dramatic. Anyone looking at that plot and considering a launch at 31 degrees would stop and think twice. This is a case in which we do not need a sophisticated mathematical model in order to make a decision[2].

**Example 2**

Here is a scatter plot of the January snowfall total (in inches), Y, vs. the average temperature for January, X, in Columbus, OH for the years 1974-1989[3]. Each point in the plot represents one year. We quickly notice that there is little snow in warm Januaries. That is, as average temperature goes up, snowfall total tends to go down. Moreover, the scatter plot gives us a sense of how strong this relationship is. We can also see that the January snowfall total for Columbus ranged from just a couple of inches up to about 35 inches, while the distribution of monthly average temperature centered around 25 degrees.

[2]For a more detailed statistical analysis, see S.R. Dalal, E.B. Fowlkes, and B. Hoadley (1989), "Risk Analysis of the Space Shuttle: Pre-Challenger Prediction of Failure," *J. Amer. Statist. Assoc.*, **84**:945-957.

[3]Data from *Climatological Data, Ohio* monthly reports. The raw data are presented on page 61.

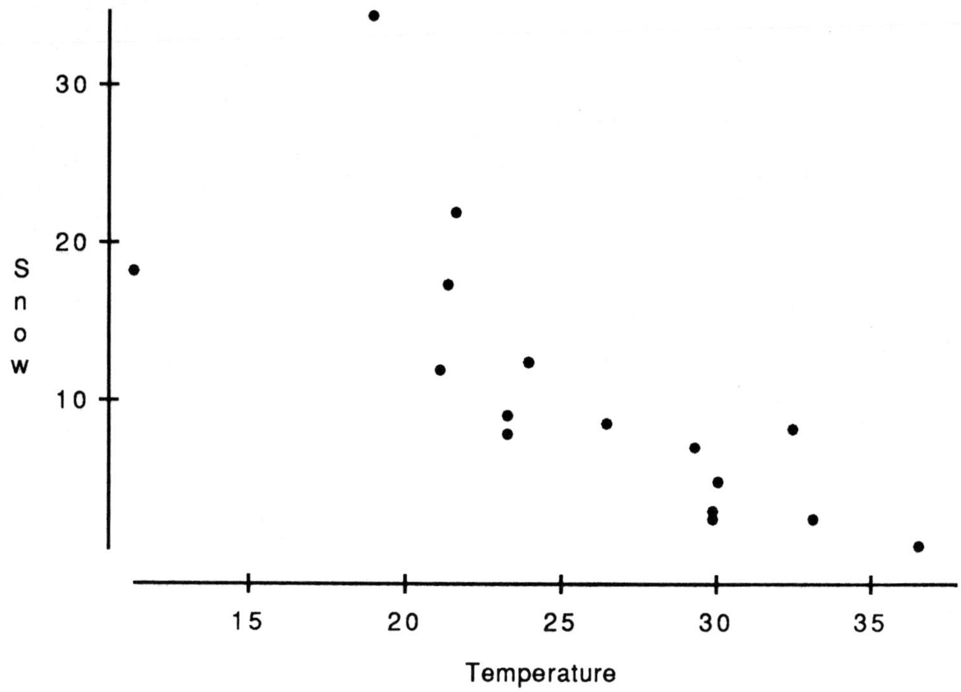


**Example 3**

The following scatter plot shows the relationship between neonatal body weight in grams, Y, and gestation time in days, X, for 23 species of mammals. The data are given in Table 4.1 below[4].

These data include two groups - primates and carnivores. To differentiate between the groups on the graph, the primates are shown with dots and the carnivores are shown with the plotting symbol x. Note that the relationship between X and Y is not linear, particularly for the primates.

[4]From G.A. Sacher and E.F. Staffeldt (1974), "Relation of Gestation Time to Brain Weight for Placental Mammals: Implications for the Theory of Vertebrate Growth," *The American Naturalist*, **108**:593-615. Copyright © 1974 by The University of Chicago; used by permission.

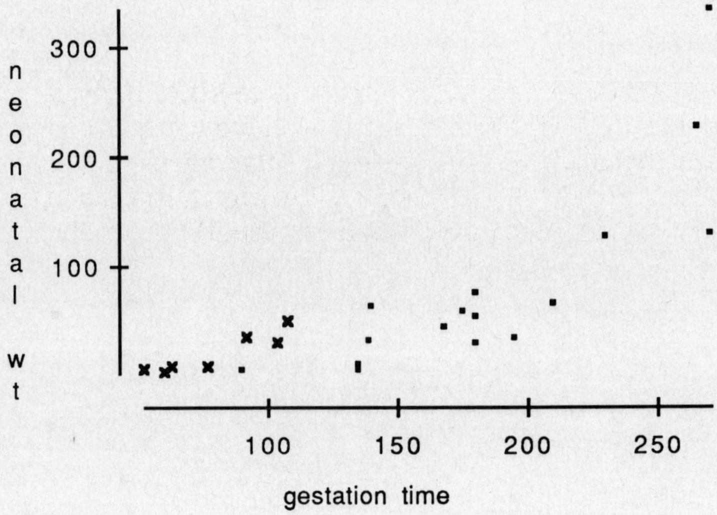


**Table 4.1**

| | Species | Neonatal body weight (grams) | Gestation time (days) |
|---|---|---|---|
| Primates | Ring-tailed lemur | 8.78 | 135 |
| | Slow loris | 4.0 | 90 |
| | Thick-tailed bushbaby | 4.0 | 135 |
| | Mantled howler monkey | 30.8 | 139 |
| | White-faced capuchin | 29 | 180 |
| | Spider monkey | 64 | 140 |
| | Rhesus monkey | 58 | 175 |
| | Baboon | 75 | 180 |
| | Guinea baboon | 53 | 180 |
| | Vervet monkey | 33.5 | 195 |
| | Dusky leaf monkey | 43 | 168 |
| | Lar gibbon | 65 | 210 |
| | Orang-utan | 129 | 270 |
| | Chimpanzee | 128 | 230 |
| | Gorilla | 227 | 265 |
| | Human | 335 | 270 |
| Carnivores | Dog (beagle) | 6.8 | 63 |
| | Red fox | 3.82 | 52 |
| | Kinkajou | 7.73 | 77 |
| | Badger | 3.34 | 60 |
| | Leopard | 35.3 | 92 |
| | Lion | 49.5 | 108 |
| | Tiger | 30 | 104 |

## 4.2 Correlation

There is considerable variability from state to state in *college%*, the percentage of adults age 25 years and older who have completed at least four years of college. Personal income per capita, *percap$*, also varies from state to state and is related to *college%*. Here is a scatter plot showing the relationship between *college%*, X, and *percap$*, Y, based on data from the 1980 census.

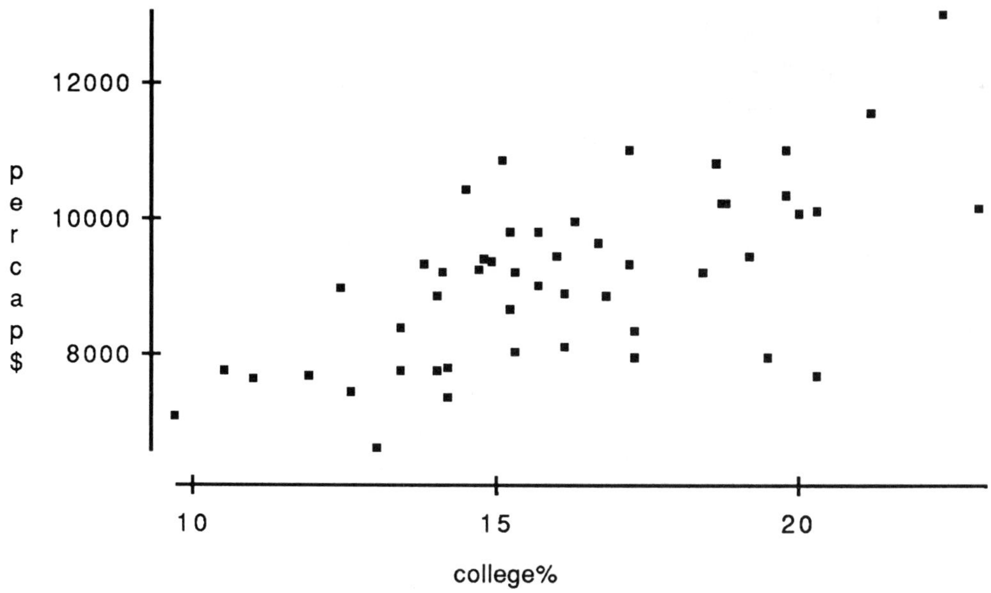


Each point in the scatter plot represents a state. For example, Alaska has a *college%* value of 22.4 and a per capita income of $13,007; the highest point on the scatter plot represents Alaska.

Note that as *college%* increases, so does *percap$*. This relationship is fairly strong, although it is by no means a perfect relationship, for at each level of *college%* there is variability in *percap$*.

A measure of the strength of the relationship between these two variables is given by the *correlation coefficient,* r=.646. As you look at the scatter plot between X and Y you can see an upward trend that seems to follow a straight line. (It does not matter which variable plays the role of X and which is Y when considering correlation.) The correlation coefficient measures how tightly the points in the scatter plot cluster about this line. If there were a perfect (positive) linear association between X and Y, then r would be 1, as in the following scatter plot:

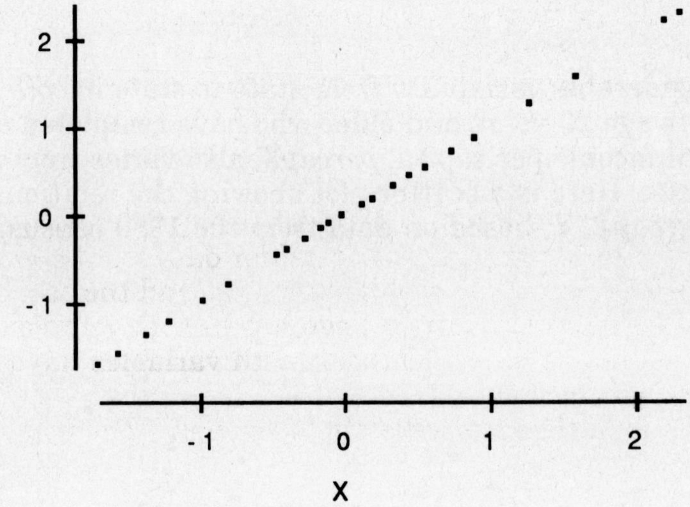


If there were a perfect linear association between X and Y, but the slope of the line were negative, then r would be -1:

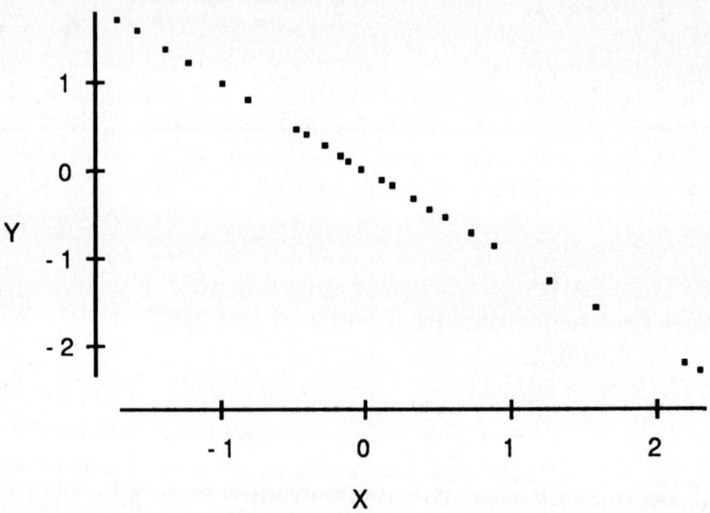


If there were no linear relationship between X and Y, then r would be 0:

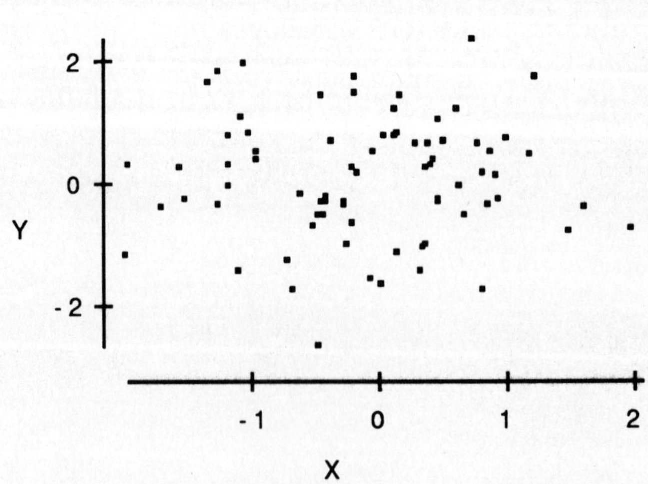

The correlation coefficient measures clustering; the tighter the clustering, the closer r is to 1 (for upward trends) or -1 (for downward trends). For the Space Shuttle data, there is a downward trend in the plot, so r is negative. The clustering is moderate; r=-.561. The clustering is a bit tighter in the Columbus weather data. The correlation between snowfall and average January temperature is -.754.

Note that a large correlation coefficient does <u>not</u> imply that changes in X cause changes in Y. For example, if we consider data for the past 20 years there is a large correlation between the cost of a new car and the cost of a college education. This does not mean that college has become more expensive because cars have become more expensive. Rather, both variables have been increasing over the years due to inflation.

Compare the following scatter plots to see how r changes as the level of clustering in the data changes.

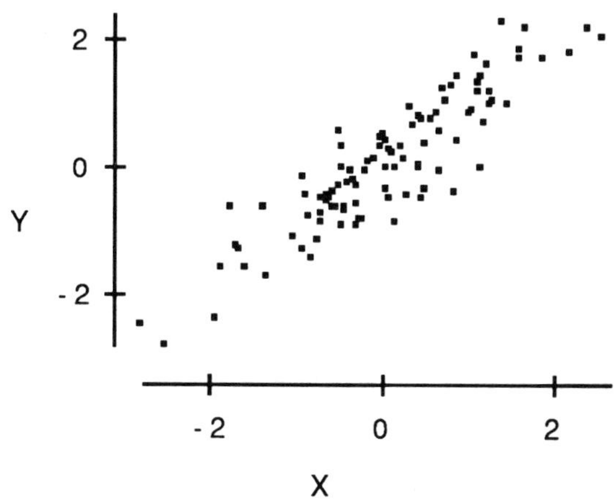


r=.9

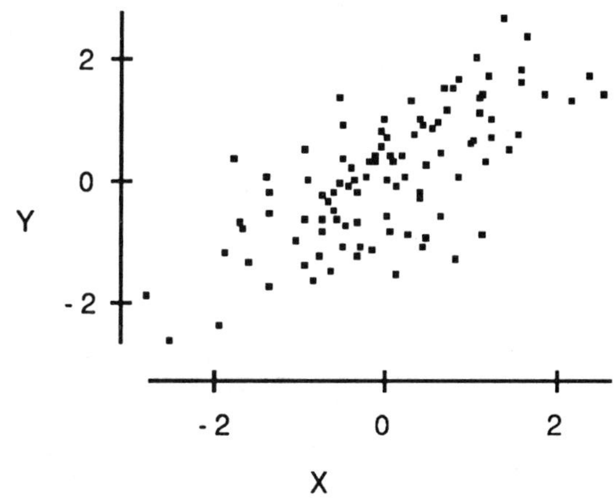


r=.7

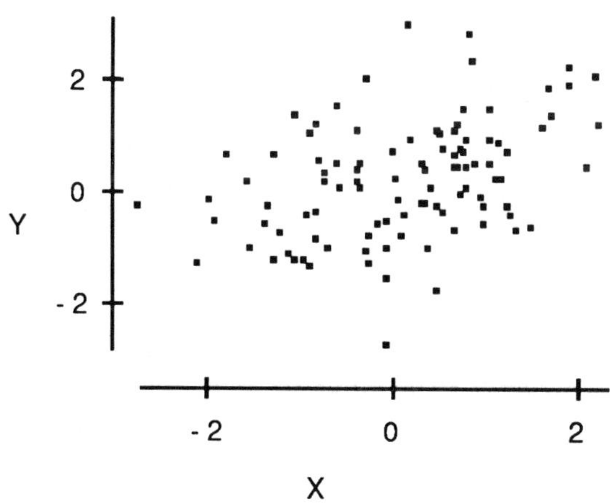


r=.4

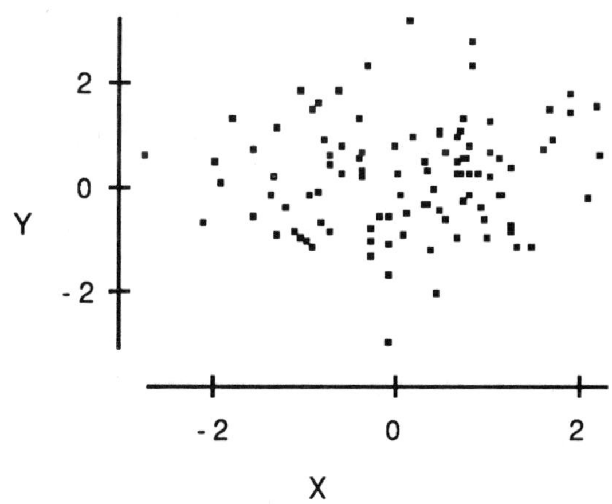


r=.1

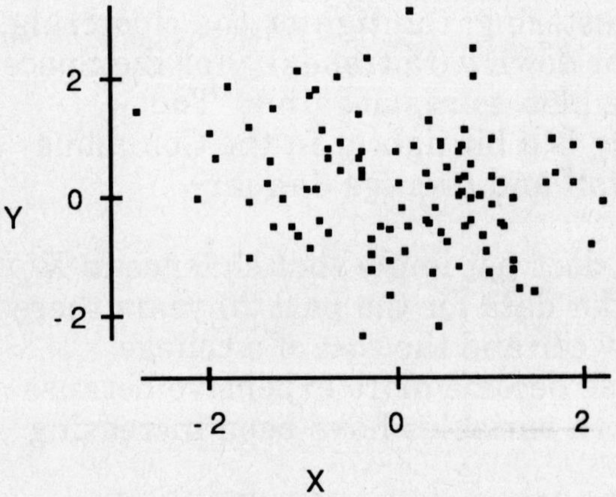


r=-.2

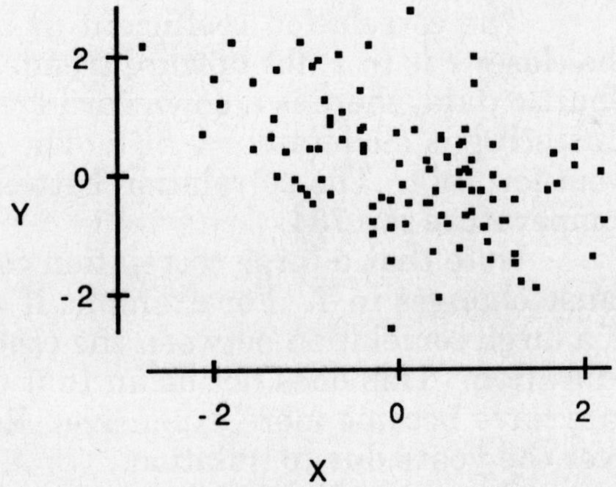


r=-.5

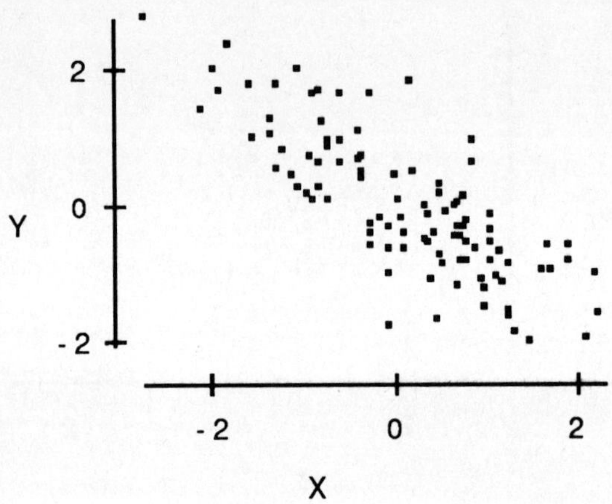


r=-.8

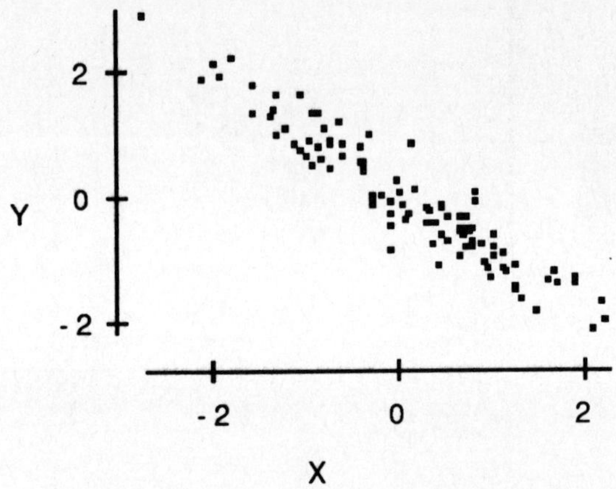


r=-.95

Note that r measures clustering, not the steepness of the slope between Y and X. Changing the scale of one of the variables changes the slope, but does not affect r. Changing the size or shape of the scatter plot will change the appearance of the scatter cloud of points, but does not change r. For example, r is .8 in each of the following plots, which show the same set of points.

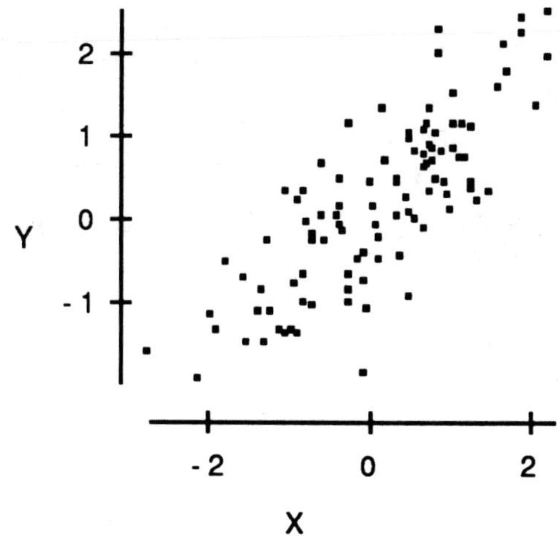
2
1
0
-1
Y
-2
0
2
X

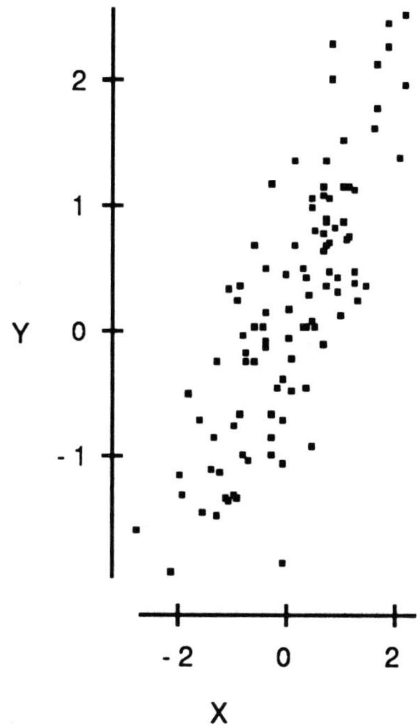
2
1
0
-1
Y
-2
0
2
X

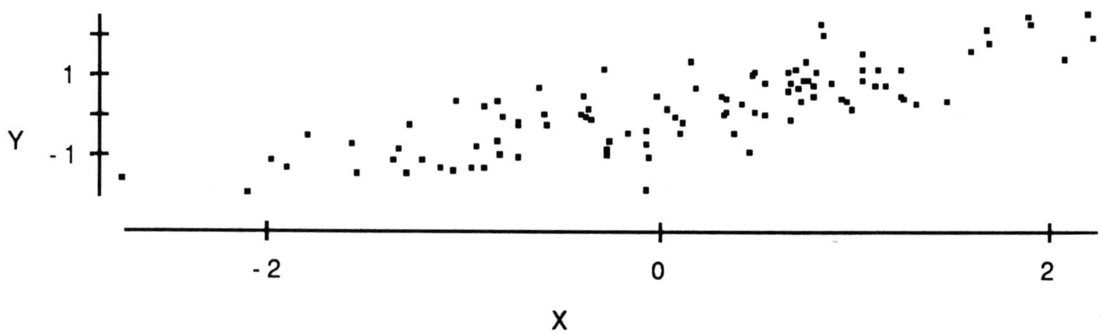
1
Y
-1
-2
0
2
X

There are two features of the correlation coefficient that make it a convenient summary of the linear relationship between two variables:

1) r does not depend on the units in which X and Y are measured, so changing the units of X or of Y (e.g., measuring Y in yards or meters, rather than in feet) has no effect on r.

2) $-1 \leq r \leq 1$ always, with r = 1 or r = -1 corresponding to perfect linear association and r = 0 corresponding to the absence of any linear association.

Note that r only measures linear association between variables. The data in the following plot show a vivid pattern, but there is no <u>linear</u> trend between X and Y, so r=0.

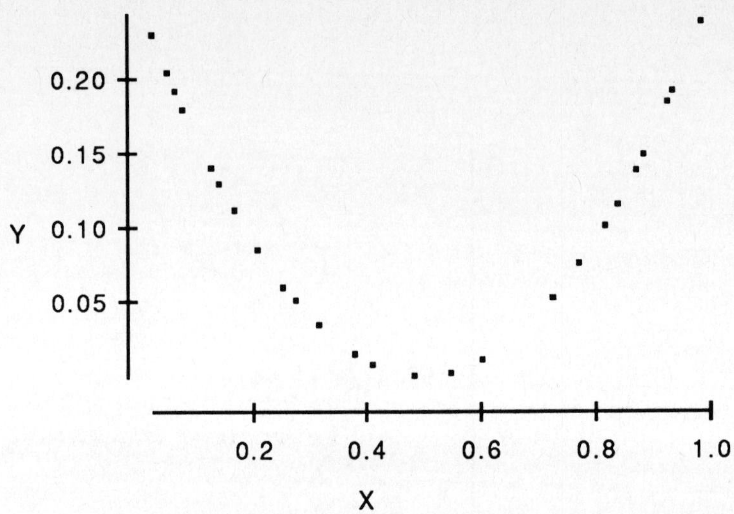

4.3 Resistant lines

There are many ways to fit a line through a scatter cloud of points. The most widely used technique is the method of least-squares regression, discussed in Section 4.4. Here we will consider a more simple and robust method.

Fitting a resistant line.

Many libraries have gates that automatically count persons as they leave the building, thus making it easy to find out how many persons used the library in a given year. (Of course, someone who enters, leaves, and enters again is counted twice.) The following scatter plot shows the relationship between gate count at the main library for one year, Y, and enrollment, X, for a sample of 17 colleges[1].

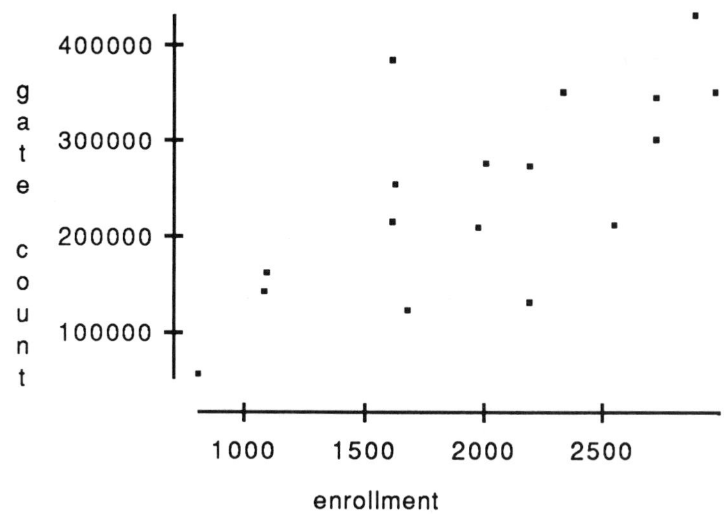


The data are given in Table 4.2, sorted from smallest to largest value on enrollment.

**Table 4.2**

| Enrollment X | Gate count Y | Enrollment X | Gate count Y |
|---|---|---|---|
| 810 | 54738 | 2337 | 350000 |
| 1086 | 141248 | 2555 | 211311 |
| 1103 | 160774 | 2729 | 298766 |
| 1621 | 214342 | 2729 | 344700 |
| 1622 | 383154 | 2893 | 430858 |
| 1627 | 254116 | 2980 | 350000 |
| 1682 | 122610 | | |
| 1979 | 209310 | | |
| 2007 | 274691 | | |
| 2193 | 272376 | | |
| 2200 | 130000 | | |

[1] These data are from a survey of liberal arts college libraries, conducted in 1988 by Alison Ricker and Jeff Witmer.

To summarize the relationship between X and Y we fit a straight line through the scatter plot. To fit a line that is resistant to outlying (unusual) values in X or in Y we proceed in steps[2].

1. Ignoring Y (gate count), we split the data into three regions according to X, creating a low-X region, a middle-X region, and a high-X region; see the plot below.

We want to have nearly one-third of the points in each of the three regions. Since we have 17 points, we cannot get a perfectly even split. We put 6 points in the first region, 5 points in the middle, and 6 points in the last region.

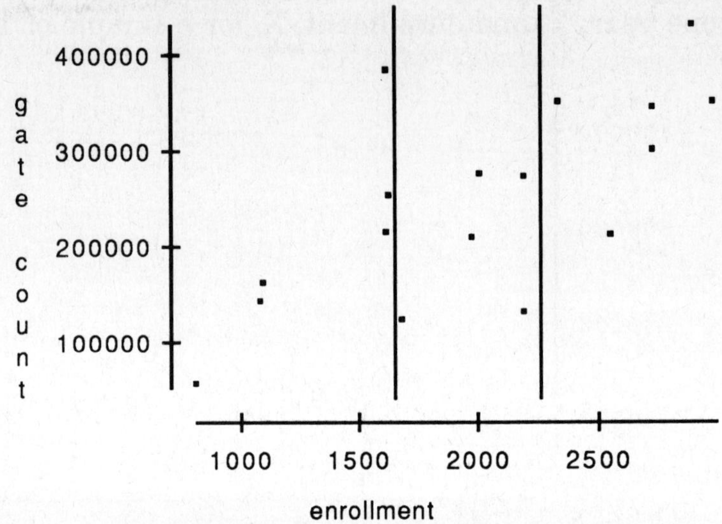


2. For the low-X region, ignore the pairing of X and Y values and, treating the X-values and Y-values separately,
 (a) find the median of the X-values,
 (b) find the median of the Y-values,
 (c) pair the value from (a) with the value from (b) and plot this new summary point, denoted $(X_L,Y_L)$, on the scatter plot, using a new plotting symbol.

In our example, the low-X region contains 6 points. The median of the 6 X-values is 1362; the median of the 6 Y-values is 187558. The point $(X_L,Y_L)$=(1362,187558) and is plotted below with a triangle.

3. Repeat Step 2 for the middle-X region; call the resulting point $(X_M,Y_M)$. In the middle-X region there are 5 points. The median of the 5 X-values is 2007. The median of the 5 Y-values is 209310. (Note that in the set of 5 data points, the X-value of 2007 is paired with a Y-value of 274691, which is <u>not</u> $Y_M$.) Thus, $(X_M,Y_M)$ is (2007,209310).

[2] There are many methods of fitting resistant lines to data. The method presented here is often called the "median fit" technique.

4. Repeat Step 2 for the high-X region; call the resulting point $(X_H, Y_H)$. The median of the 6 X-values in the high-X region is 2729; the median of the 6 Y-values is 347350. Thus, $(X_H, Y_H)$ is (2729,347350).

5. Look at the three summary points plotted on the scatter plot. These three points should come reasonably close to being on a straight line. If not, then the relationship between X and Y is not linear and we should not use a line to summarize the relationship.

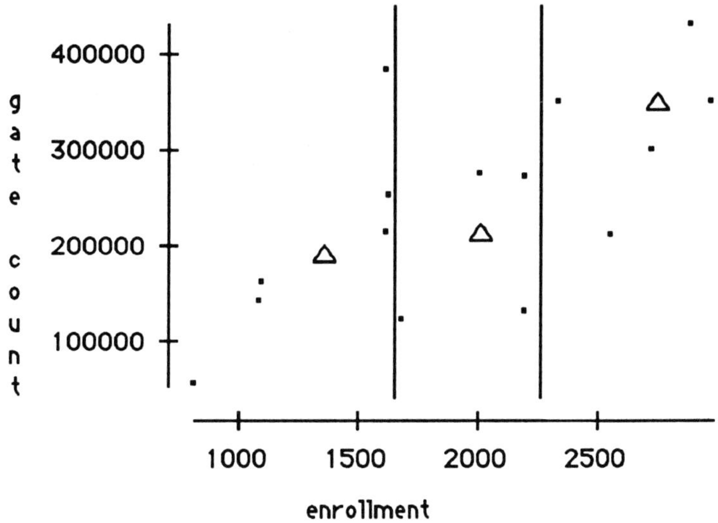


6. The resistant line is found by connecting the summary points in the two outside regions and then sliding this line one-third of the way to the middle summary point. Thus, the slope is determined by the summary points $(X_L, Y_L)$ and $(X_H, Y_H)$ and the intercept is determined using all three summary points. More formally, the slope, b, of the resistant line is given by

$$b = \frac{Y_H - Y_L}{X_H - X_L}$$

and the intercept, a, is given by

$$a = 1/3[(Y_L + Y_M + Y_H) - b(X_L + X_M + X_H)].$$

In our example, the slope is

$$b = \frac{347350 - 187558}{2729 - 1362} = 116.9$$

and the intercept is

$$a = 1/3[(187558 + 209310 + 347350) - 116.9(1362 + 2007 + 2729)]$$
$$= 10454.$$

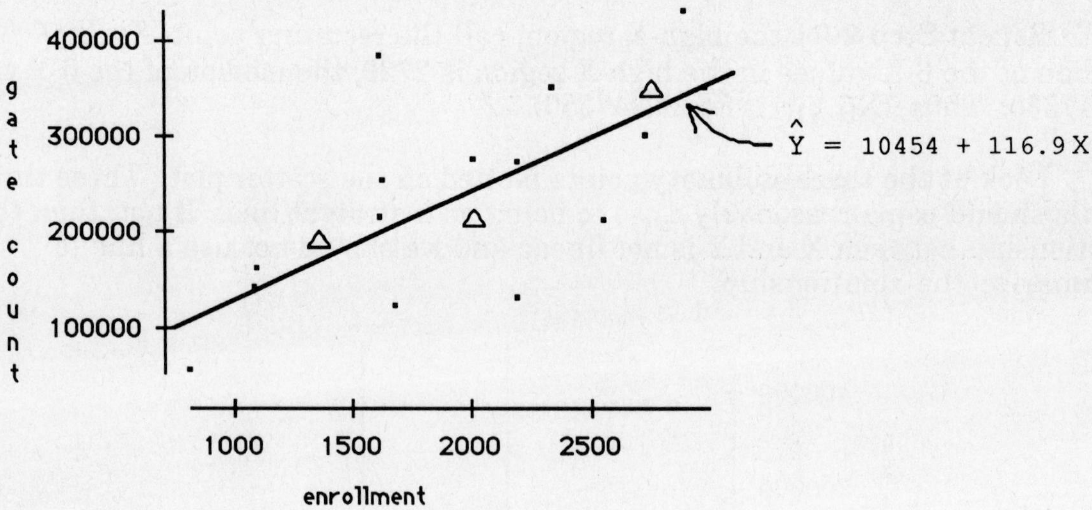


The resistant line is $\hat{Y} = 10454 + 116.9X$. The "hat" on Y here indicates that the values of Y generated by this equation (by plugging values in for X) are estimates. The line summarizes the trend in the data, indicating that a difference in enrollments, from one college to another, of 100 students corresponds to a predicted difference in gate counts at the main libraries of 116.9*100 = 11,690.

An attractive feature of a resistant line (indeed, the reason such lines are called "resistant") is that the line is not sensitive to the presence of a few unusual points in the data set. For example, the last observation in Table 4.1 is (2980,350000). Even if the value of Y here changed to from 350,000 to 500,000, the third summary point would still be (2729,347350), since it is based on medians. Thus, the line would not change.

Note: Some computer packages have routines that fit resistant lines to data. For example, the computer package MINITAB® has a command called RLINE that fits resistant lines.

Minitab is a registered trademark of Minitab, Inc.

Exercises

1. Here are the data from Example 2 of Section 4.1:

| Average Temperature X | January Snowfall Y |
|---|---|
| 33.2 | 2.3 |
| 32.5 | 8.1 |
| 24.0 | 12.4 |
| 11.4 | 18.1 |
| 19.0 | 34.4 |
| 21.4 | 17.3 |
| 29.3 | 7.0 |
| 23.3 | 7.8 |
| 21.2 | 11.8 |
| 29.9 | 2.6 |
| 23.3 | 9.0 |
| 21.7 | 21.9 |
| 30.1 | 4.8 |
| 29.9 | 2.7 |
| 26.5 | 8.4 |
| 36.6 | 0.6 |

(a) Fit a resistant line to the data. There are 16 observations here, so we would like to put 5 points in the first region, 6 in the middle, and 5 in the upper region. However, there are two ties among the X-values: there were two Januaries in which the average temperature was 23.3 and two Januaries with average temperature of 29.9. The tie at 23.3 does not matter, since both of these points would go into the middle region if they were not tied on X. The tie at 29.9 is a small problem, since we would put one of these points in the middle and one in the upper region if they were not tied. Do not split these points; rather, put both of them in the upper region (to make things as balanced as possible). Thus, put 5 points in the low-X region, 5 points in the middle, and 6 points in the upper region.

(b) Interpret the line found in part (a).

2. Fit a resistant line to the Space Shuttle data from Section 4.1.

4.4 Least-squares Regression Lines

The most widely used technique of fitting lines to data is the method of *least-squares regression.* Virtually every statistics computer package includes a least-squares procedure. As usual, we begin our discussion with an example.

**Example 1 - Per Capita Income (continued)**

Consider the per capita income data of Section 4.2. Suppose we want to learn the extent to which changes from state to state in X, *college%*, are associated changes in Y, *percap$*. Here we are thinking of Y as depending on X. We will fit a straight line of the form $Y = \beta_0 + \beta_1 X$ through the scatter plot to help understand how changes in X are related to changes in Y.

How should we choose $\beta_0$ and $\beta_1$? Any choices of $\beta_0$ (intercept) and $\beta_1$ (slope) will produce a line. To judge whether or not the line we have is good, we can measure how close the line comes to each of the points. Since we think of Y as depending on X, we will measure the distance from a point to the line in the vertical direction. This distance is called the *residual* for the point. See the diagram below.

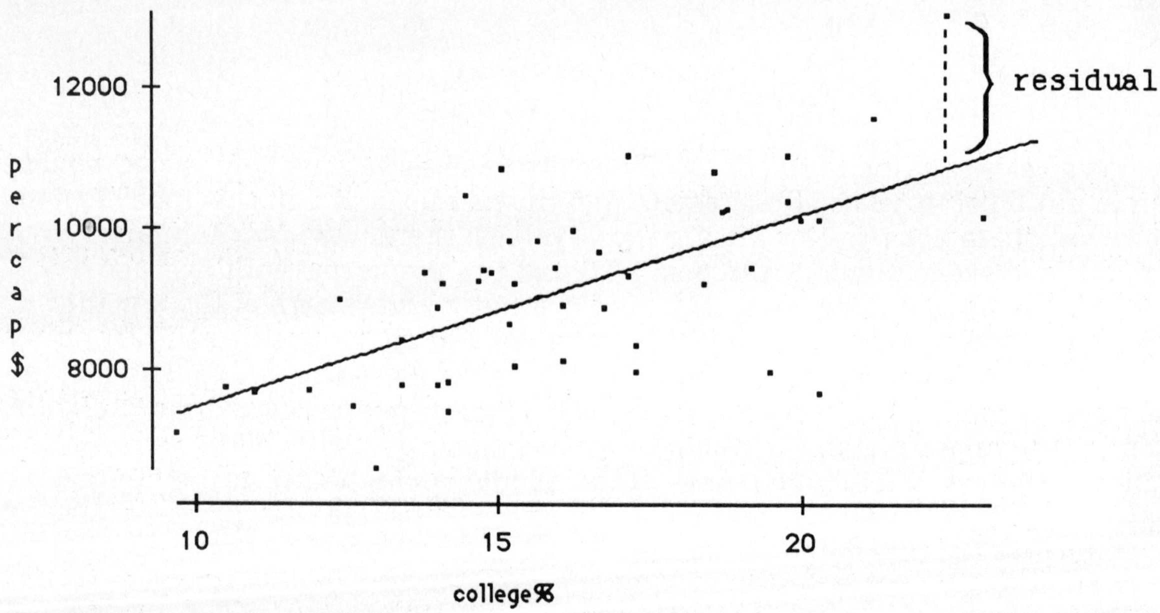


Any line that goes through the center of the scatter cloud of points will yield some positive residuals (when points are above the line) and some negative residuals (when points are below the line). If we take the residuals for each of the points, square them, and then add together these squared residuals, we obtain a measure of how well the line fits the scatter cloud of points. The method of least-squares produces the unique line that minimizes this sum of squared residuals.

There are many nice theoretical properties of least-squares lines, which we will not consider here. Also, the least-squares method is fairly easy to implement on a computer. These two features combine to make least-squares very popular with statisticians. However, unlike the method presented in Section 4.3, the least-squares method is not robust against outliers in either X or Y.

The least-squares line for the per capita income data is $\hat{Y} = 4696 + 276X$ and is plotted below.

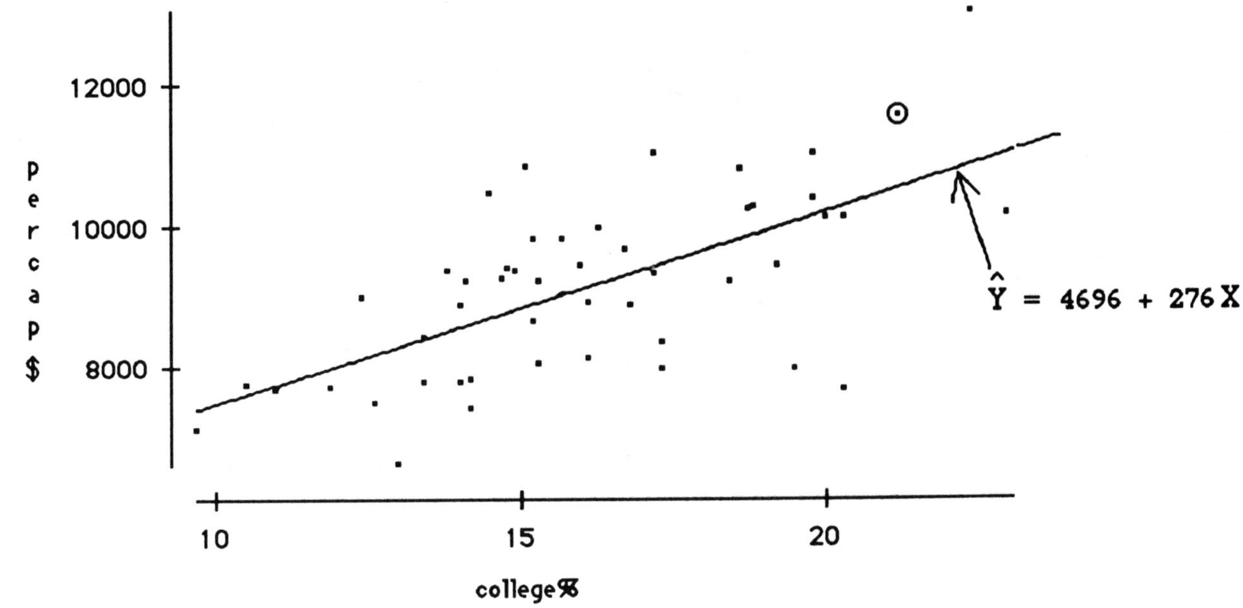


The point circled in the previous graph is for Connecticut. As of 1980, 21.2% of adults age 25 and older in Connecticut had completed four years of college (X=21.2) while the per capita income was $11,559 (Y=11,559). Using the least-squares line, we predict Y to be 4696 + 276(21.2) = 10,547. Thus, the residual for Connecticut is 11,559 - 10,547 = 1,012, which is positive since the dot for Connecticut is above the least-squares line. The squared residual for Connecticut is $1{,}012^2$, which equals 1,024,144.

Likewise, we can find the residuals for each of other states. Squaring these residuals, which are sometimes called "errors," and then adding yields a total, called the *Sum of Squared Errors*, or *SSE*, of 47,436,964. This total is as small as it can be, since it is the result of using the method of least-squares. Any other method of fitting a straight line to the data would yield a larger SSE.

We can interpret the slope, 276, as follows: going from state to state, as the percent of adults who have completed four years of college goes up by 1%, per capita income is estimated to go up by $276.

Another way to think about the regression line is to consider the *percap$* values of states for which *college%* is roughly 14%. These states are highlighted below.

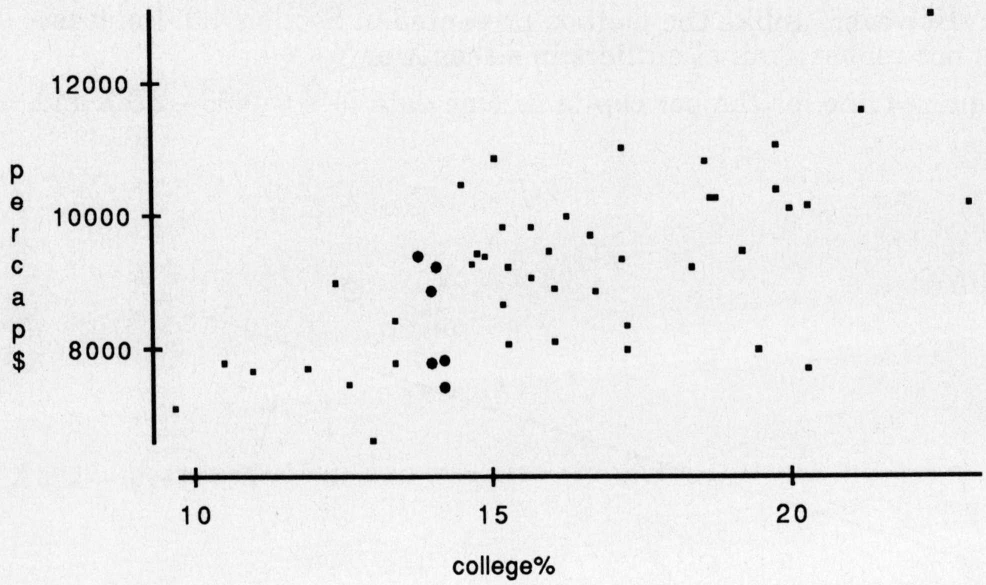


The average *percap$* for these 6 states is $8397.

Likewise, we can consider states for which *college%* is roughly 16% and find the average *percap$* of the 4 states in this vertical strip ($9094).

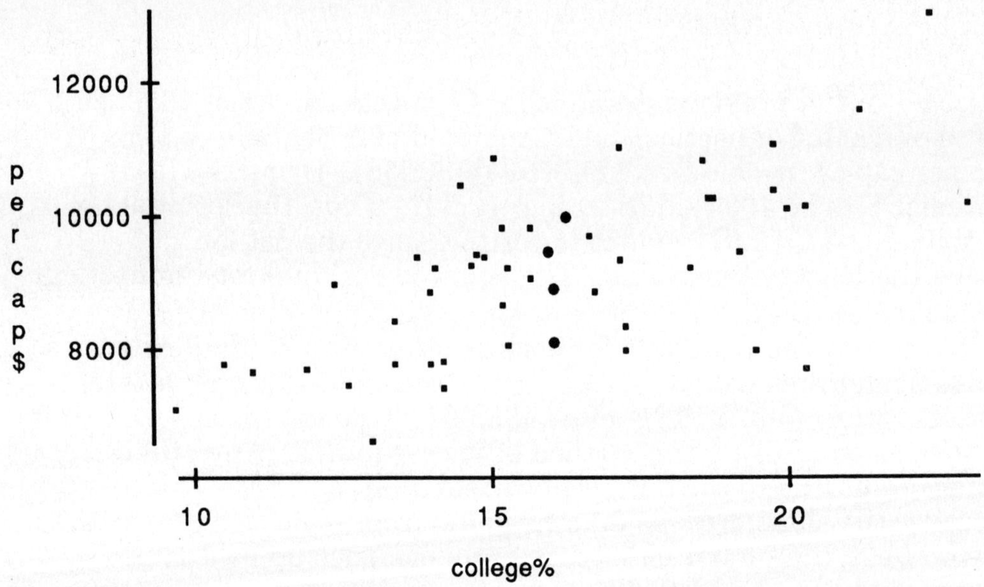


If we do this for other values of *college%* we get the *graph of averages*, which is a graph of the average value of Y for fixed values of X.

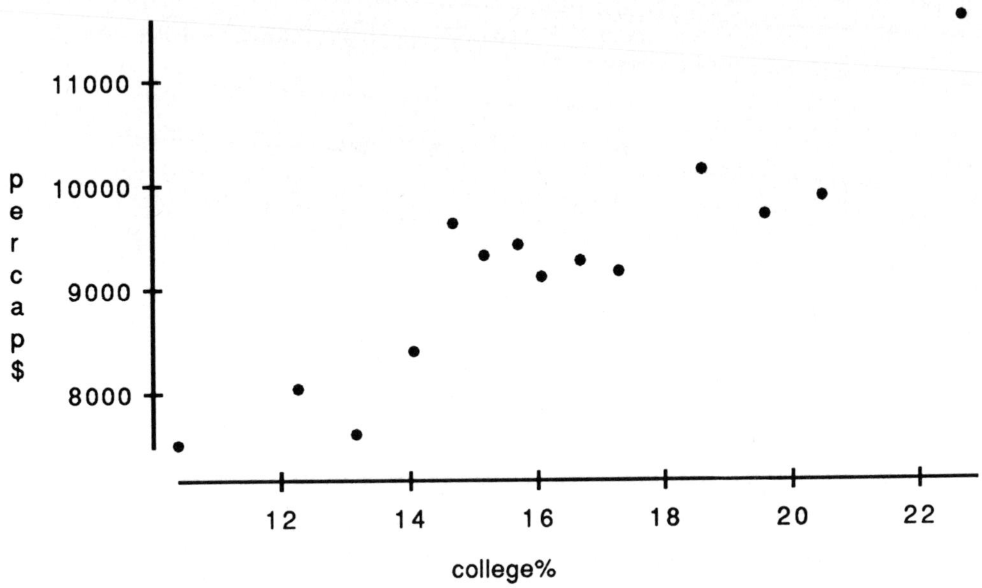


Note the upward linear trend in this plot. The line suggested by this plot is the least-squares regression line. That is, the least-squares regression line is a smoothed ("linearized") version of the graph of averages.

**Example 2 - The Space Shuttle (continued)**

As another example, here is a scatter plot of the Space Shuttle data with the least-squares line, $\hat{Y} = 4.79 - 0.0627X$ included.

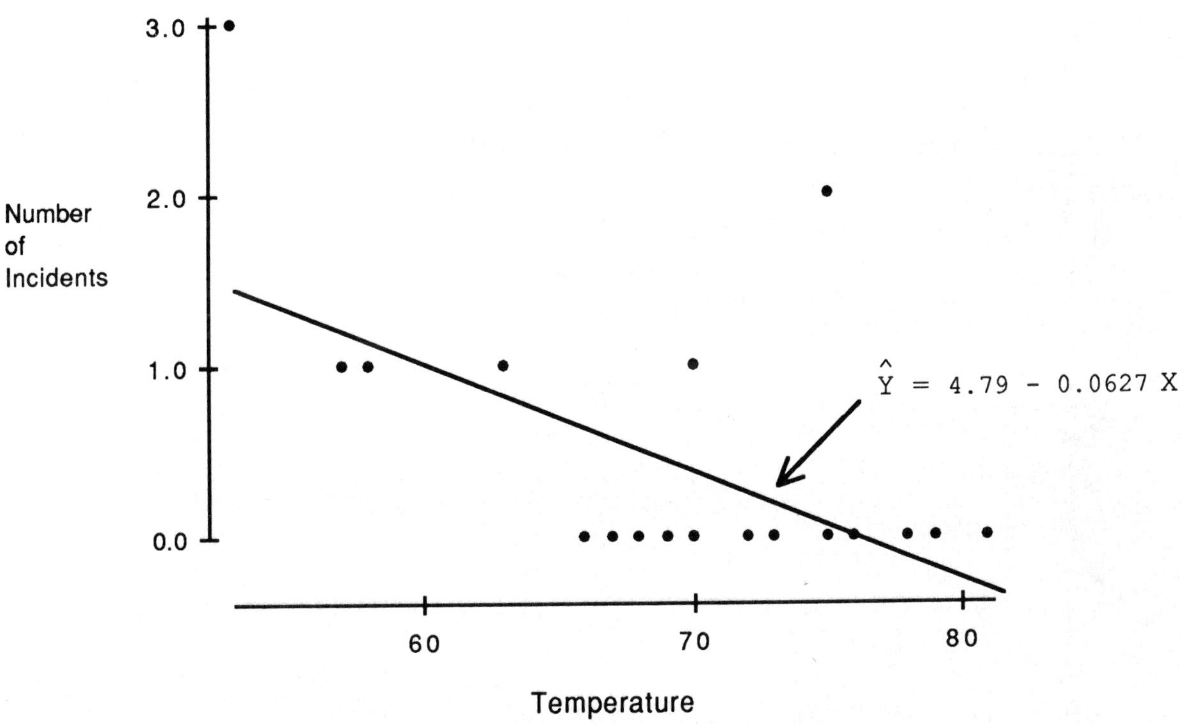

The slope of -0.0627 summarizes the trend that as temperature goes up the number of incidents of O-ring thermal distress goes down.

Note that the line goes below zero when temperature, X, is above 76. That is, the least-squares model predicts a negative number of damage incidents when the temperature is very high. Of course, the number of damage incidents cannot be lower than zero. This points up a limitation in the model. The general trend in the data, as captured by the least-squares line, does not hold when temperature is very high.

### 4.4.1 $r^2$

There is a good deal of variability from state to state in per capita income and we have seen that per capita income is linearly related to the variable *college%*. That is, differences between the states in the percentage of adults who have completed four years of college accounts for some of the variability among states in per capita income - states with a large *college%* tend to have a large *percap$*. We can interpret $r^2$, the square of the correlation coefficient, as the percentage of the variability in Y (*percap$*) that is accounted for by variation in X (*college%*). Since $r^2=.646^2=.42$, we say that 42% of the variability from state to state in per capita income can be accounted for by considering the linear relationship between per capita income and *college%*.

Note that $r^2$ is always between zero and one. Also, if r=1 or r=-1, then $r^2$ equals 1, so that 100% of the variability in Y is accounted for by X (which makes sense, since r=1 or r=-1 means that the points in the scatter cloud fall exactly on a straight line and we can make perfect predictions about Y if we know X). If r=0, then $r^2=0$ and X accounts for none of the variability in Y.

Here we are measuring variability in terms of the standard deviation, as opposed to, say, the interquartile range. The standard deviation of the <u>residuals</u> (the extent to which *percap$* in a state exceeds or falls short of what the regression model predicts) is $\sqrt{1 - .42}*1276.5 = 972$. (The variance of *percap$* is $1276.5^2 = 1629560$, whereas the variance of the residuals is $.58*1629560 = 945145$.)

4.5 The Regression Effect

Here is a scatter plot of data collected during the 1989 major league baseball season. In the plot, X is a player's batting average for games played before the mid-season All-Star Game break; Y is the player's batting average for games played after the All-Star break. Each point in the scatter plot represents one of the Minnesota Twins.

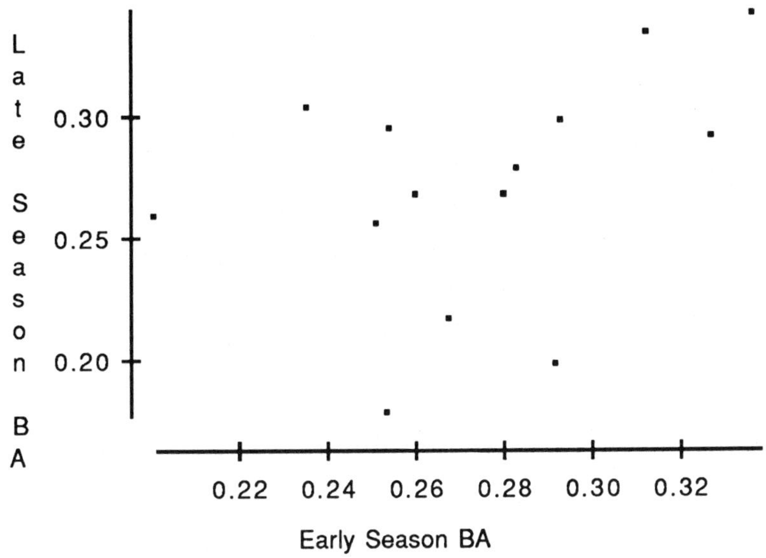


The correlation between X and Y is .398. The least-squares line is $\hat{Y} = .128 + .516X$. The Twins batted .274 as a team before the All-Star break and .279 after the break, and the variability in the data stayed the same, so we might expect that each player would have, roughly, the same batting average after the break as he had before the break. This would mean that the slope of the line would be 1, rather than .516.

For reference purposes, both the least-squares line - the solid line - and a line with slope 1 - the dashed line - are included on the following graph. The dashed line is sometimes referred to as "the SD line." In general, this line goes through the point (average of X, average of Y) and has slope equal to (standard deviation of Y)/(standard deviation of X). In our example, it goes through the point (.274,.279) and has slope 1.

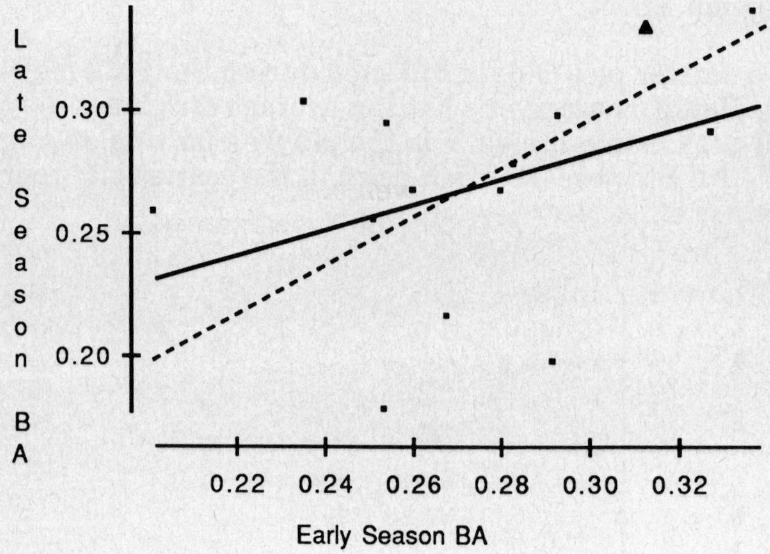


In fact, the least-squares line shows that players who did well before the break tended to regress a bit after the break. On the other hand, players who did poorly before the break tended to do somewhat better after the break. This phenomenon is know as the *regression effect*. Least-squares lines are often referred to as regression lines.

The regression effect is among the most common and misunderstood of statistical phenomena. Any linear association between two variables in which the correlation is not perfect (i.e., r is not 1 or -1) produces the regression effect: points that are one standard deviation above average in terms of X will tend to be only r standard deviations above average in terms of Y. Thus, the solid (regression) line is not as steep as the dashed line in the previous graph. Likewise, points that are one standard deviation below average when viewed in terms of X will tend to be r standard deviations below average when viewed in terms of Y.

Of course, it is possible for a point to be farther above average in terms of Y than in terms of X - Brian Harper batted .313 before and .333 after the break; this point is shown as a triangle on the graph. But this is less common than what we usually see: points (baseball players) above average in X (who hit well during the first half of the season) tend to regress somewhat (tend to cool off a bit during the second half of the season). Although they usually remain above average on Y, they tend not to be as far above average on Y as on X.

The regression effect is due to the lack of perfect correlation between X and Y. The *regression fallacy* consists of constructing an ad hoc explanation for the regression effect. For example, many fans think that baseball players who do well early in the season become lazy, while those who slump early in year work harder in the second half of the season, and attribute the regression effect to such work habits, rather than to the variability that is inherent the data (and which produces the observed correlation and regression effect).

The same phenomenon is evident when one considers examination scores of students. Those who do very well on the first exam in a class tend to regress somewhat on the second exam. They tend to remain above average, but are not as far above average on the second exam as on the first exam. This is due to the fact that there is variability in the performance of the typical student.

On any given exam some students do unusually well in comparison to their overall average (they might say that they had a lucky day). On the second exam these students will tend to drop back toward the class average. Of course, some students will do well on both exams because they are outstanding students, so not everyone who earns an A on the first exam will drop back on the second exam. But as a group, those who do extremely well on the first exam tend to do well, but not as well, on the second exam. Hence the term *regression*.

At the other end of the spectrum, students who do poorly on the first exam include both those who are weak students and those who get unusually low grades (for example, one might have a headache during the exam). The weak students will tend to do poorly again on the second exam, but the others (average students who had a bad day when the first exam was given) will tend to move up toward the class average. Thus, as a group, those who do extremely poorly on the first exam tend to do poorly, but not as poorly, on the second exam.

The regression effect would not be present if students did not vary from day to day, that is, if knowing a students score on the first exam allowed us to make a perfect prediction of the students score on the second exam. But students do vary from day to day and from test to test, so the correlation coefficient is not 1, and we see the regression effect. An example of the regression fallacy is (incorrectly) reasoning that students who do extremely well on the first exam do less well on the second exam because they become lazy or complacent.

5.1 Residual Plots

Fitting a line of the form $Y = \beta_0 + \beta_1 X$ to some data involves using the model $Y_i = \beta_0 + \beta_1 X_i + \varepsilon_i$, where $Y_i$ denotes the value of the response variable Y for the ith observation, $X_i$ denotes the value of the predictor variable X for the ith observation, and $\varepsilon_i$ represents a random error component. We typically make several assumptions: that the model is appropriate (i.e., there is a linear trend in the data), that the variance of the response, Y, is constant (i.e., the variability in Y does not change as X changes), and that the errors (the $\varepsilon_i$'s) are normally distributed. A *residual plot* is an excellent diagnostic tool for checking these assumptions.

Consider a fitted regression model $\hat{Y}_i = \hat{\beta}_0 + \hat{\beta}_1 X_i$. For each point in the data set, we define the residual, $e_i$, as $e_i = Y_i - \hat{Y}_i$. A residual plot is a plot of the residuals against the fitted (predicted) values for Y. That is, a residual plot is a scatter plot of the points $(\hat{Y}_i, e_i)$, with the residuals, $e_i$, plotted in the vertical direction and the fitted values, $\hat{Y}_i$, plotted in the horizontal direction[1].)

**Example 1 - Per Capita Income (continued)**

Here is a residual plot from fitting the regression model $Y_i = b_0 + b_1 X_i + e_i$ to the per capita income data of Section 4.2.

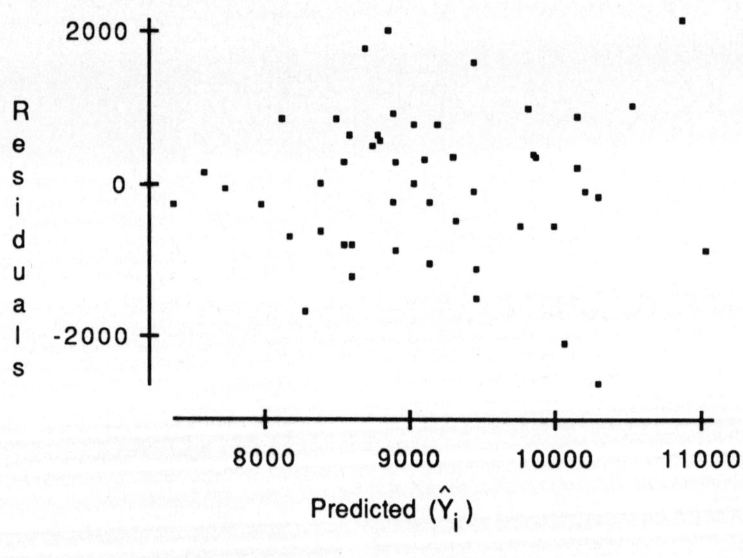


When we look at a residual plot we hope to see a random scatter of points without any pattern, as is the case here. This indicates that the linear model is appropriate and that the variance of Y is constant.

---

[1]It is standard practice to plot residuals against predicted $\hat{Y}$'s, rather than against the observed Y's, because the residuals and the Y's are usually correlated, but the residuals and the $\hat{Y}$'s are not.

**Example 2 - Neonatal Animal Weight**

The following residual plot is from the regression of neonatal body weight in grams, Y, on gestation time in days, X, for 23 species of mammals. The data and a scatter plot were given in Section 4.1. For reference purposes, a horizontal line at 0 is included in the residual plot.

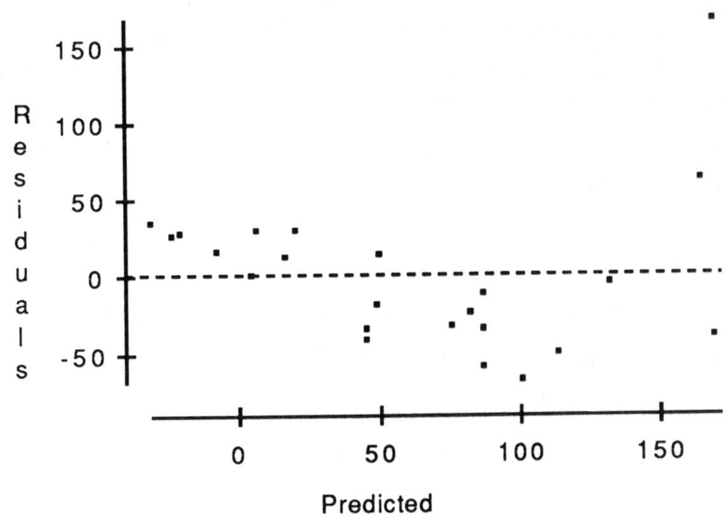


The U-shape of the residual plot indicates that the relationship between X and Y is not linear. When the predicted values are small, the residuals are positive; they are negative in the middle of the residual plot; they become large as the predicted values become large. Thus, the model $Y_i = \beta_0 + \beta_1 X_i + \varepsilon_i$ is not appropriate. Rather, a quadratic model ($Y_i = \beta_0 + \beta_1 X_i + \beta_2 X_i^2 + \varepsilon_i$) should be considered.

**Example 3 - Library Science Budget**

The next example comes from a study of liberal arts college libraries mentioned in Section 4.3. Here Y is the size of a college library's science materials budget and X is the number of science faculty at the college. The scatter plot shows a linear trend between these variables.

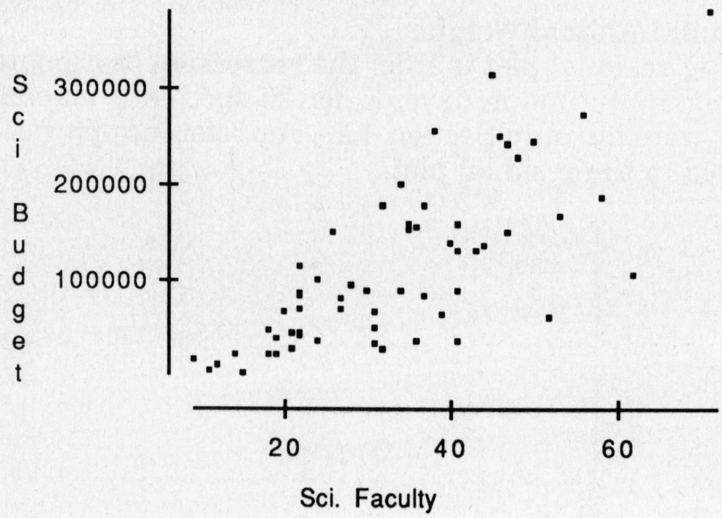


However, the variability in budgets increases as the number of faculty increases. This pattern can be seen in the scatter plot, but it shows up more clearly in the residual plot, where it is indicated by the "megaphone" shape of the residuals. If a regression model is fit to these data it will fit the points with small values of X better than it will fit the other points. These small-X points will have undue influence in determining the fitted model. A megaphone pattern calls for a transformation (see Section 5.2).

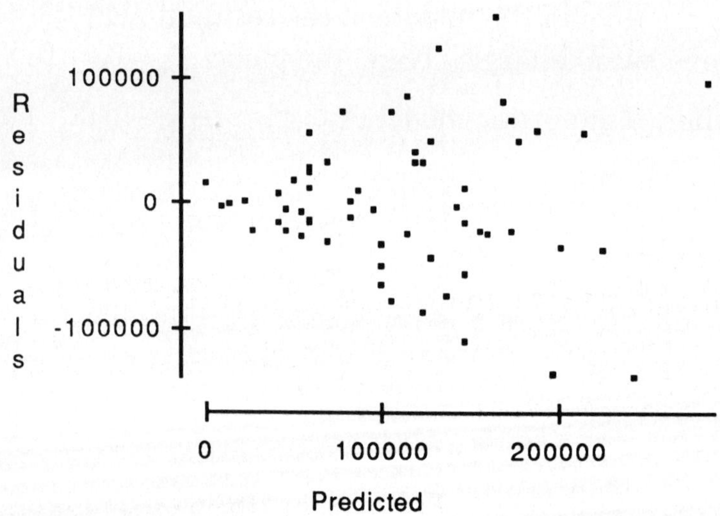


**Example 4 - Class Size and Rating**

The following plots are from a study of the relationship between class size, X, and average student rating of the class, Y, at Oberlin College. The scatter plot, on the left, shows that there is considerable variability in class ratings, which are related to class size. The residual plot, on the right, shows that the variability in the residuals decreases as the predicted class rating increases.

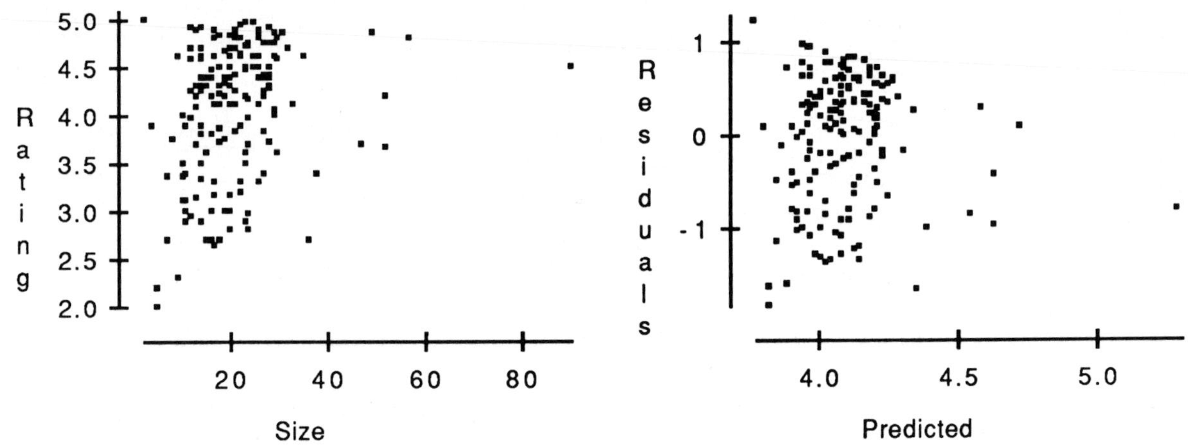


The following plot of the residuals versus class size is quite similar to the previous residual plot and shows that the variability in the residuals is largest for small classes.

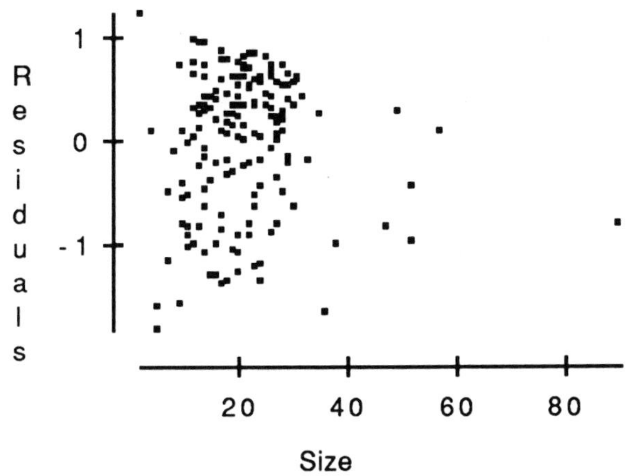


We refer to the shape of the residual plot as a "left-opening megaphone." This pattern indicates that the small group of large classes (the points on the right-hand side of the original scatter plot) have a great deal of influence when we fit the regression line to the data.

**Example 5**

We can sometimes spot an outlier more easily on a residual plot than on a scatter plot, as in the following case.

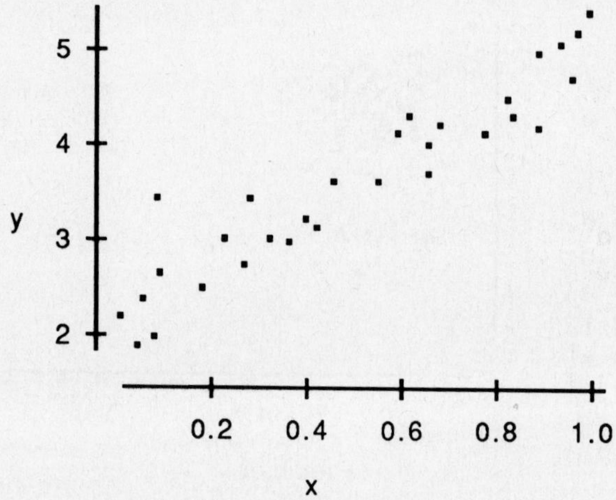


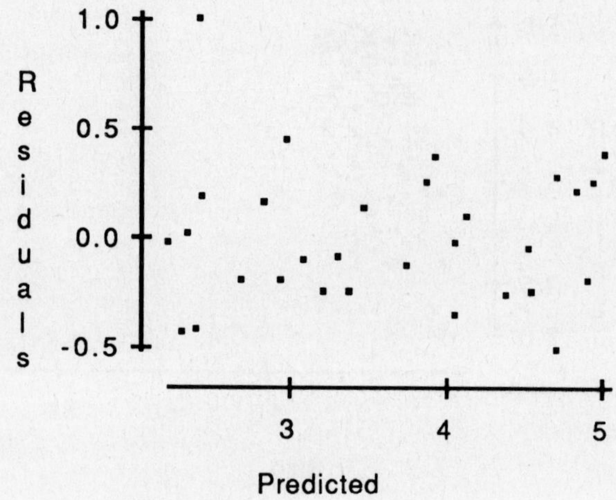


5.1.1 The Anscombe plots

In 1973, Anscombe published a collection of four hypothetical data sets that underscore the need for plots in regression[2]. Each set of data contains 11 (X,Y) points with a correlation coefficient between X and Y of .817. Moreover, the fitted regression model is $\hat{Y} = 3 + 0.5X$ for each data set. Thus, someone who fit regression models by simply tossing the data into a computer might wrongly conclude that the four data sets are equivalent. Here are scatter plots and residual plots for the four data sets.

The first data set is a good candidate for linear regression; the residual plot does not indicate any problems.

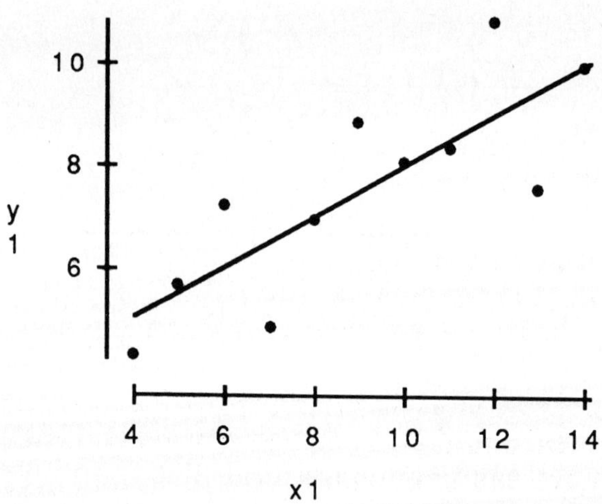


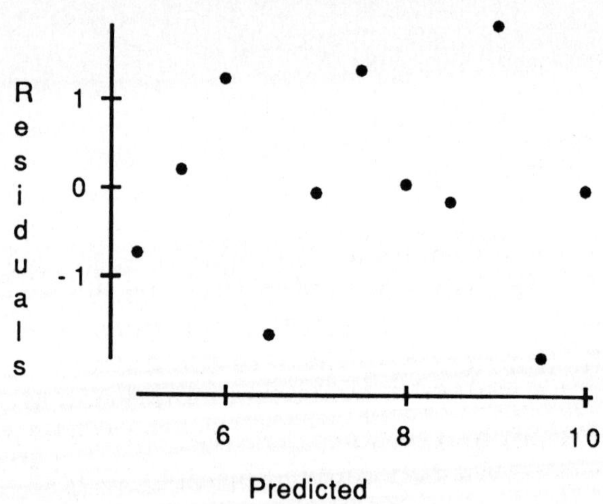


The second data set has an obvious curve, making the model $\hat{Y} = 3 + 0.5X$ clearly inappropriate. This curve shows up clearly on both the scatter plot and the residual plot.

---

[2] F.J. Anscombe (1973), "Graphs in Statistical Analysis," *Amer. Statist.*, **27**:17-21.

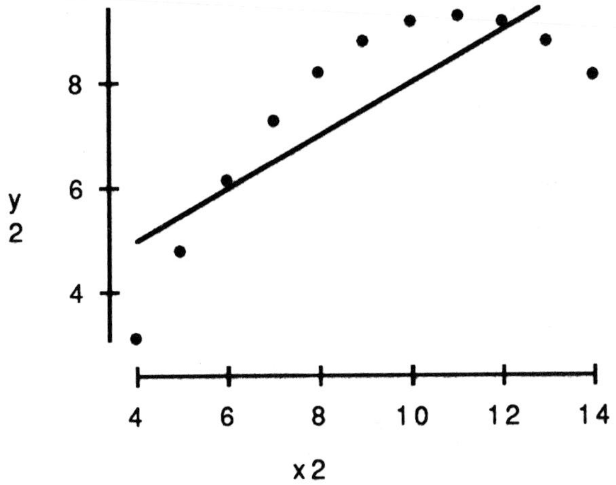


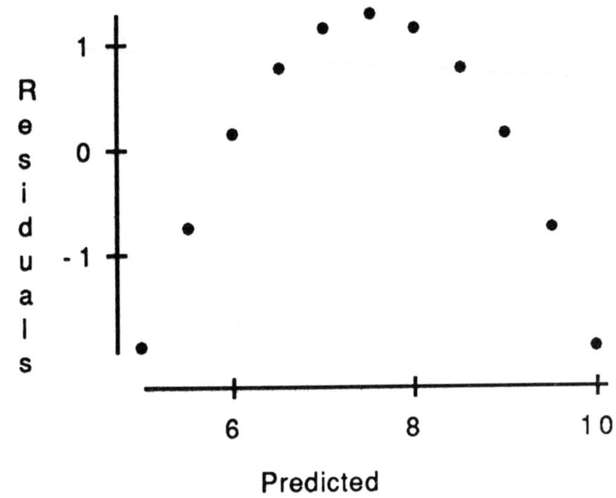


The third data set contains an outlier that has a large effect on the fitted model - if we delete this point and fit a regression model to the remaining 10 points we get $\hat{Y} = 4 + 0.346X$, which is shown as a dashed line on the plot; the model $\hat{Y} = 3 + 0.5X$ is shown as a solid line. Were these real data we would need to know how this outlier arose before we could proceed with an analysis.

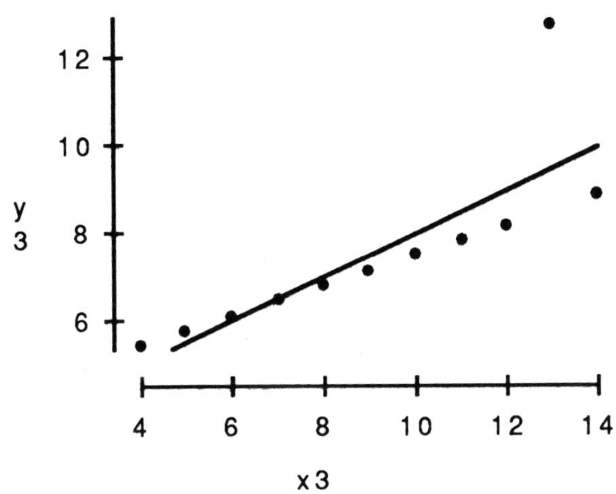


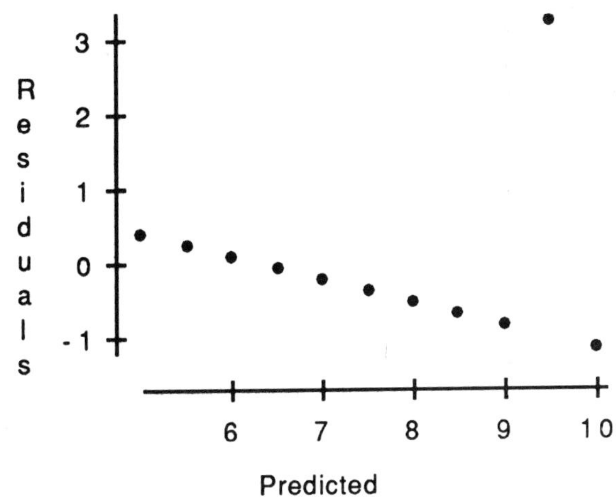


The fourth data set contains one point that is highly *influential* (see Section 5.3). The lone point determines the slope of the fitted line. We cannot be confident in a regression model that depends so heavily on a single point.

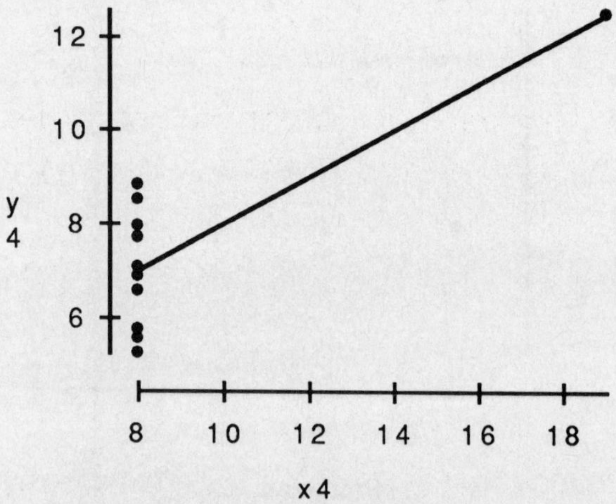


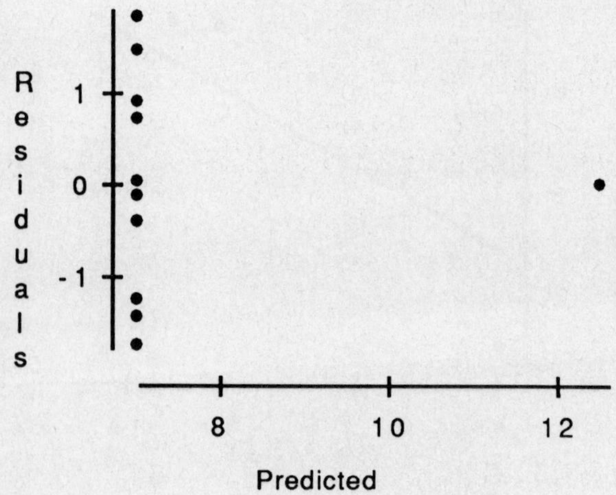


Although these data are hypothetical and show extreme situations, similar problems occur when working with real data. The moral of the story is simple: ALWAYS PLOT THE DATA.

### 5.1.2 Normal Probability Plots of Residuals

We can check the assumption that the errors in a regression model are normally distributed by examining a normal probability plot of the residuals (see Section 3.1). For example, fitting a least-squares regression line to the library gate count data of Section 4.3 yields the model $\hat{Y} = 22498 + 111.9X$. Here is a plot of the normal scores versus the residuals from this model, together with a histogram of the residuals.

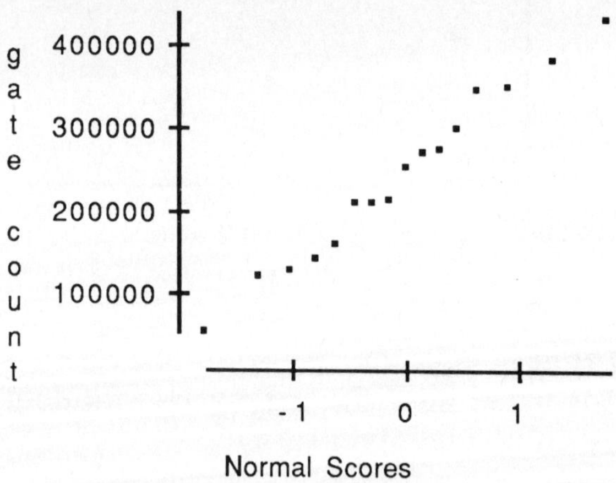


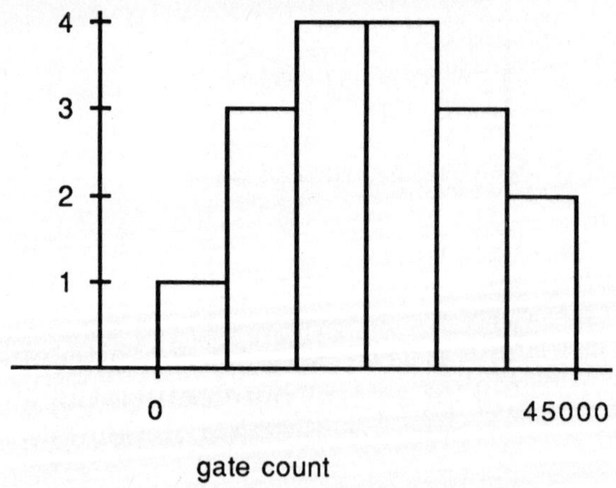


These plots support the assumption that the errors are normally distributed.

**EXAMPLE 1**

Here is a scatter plot that shows how the population of Florida has grown over the years 1830 to 1990[1].

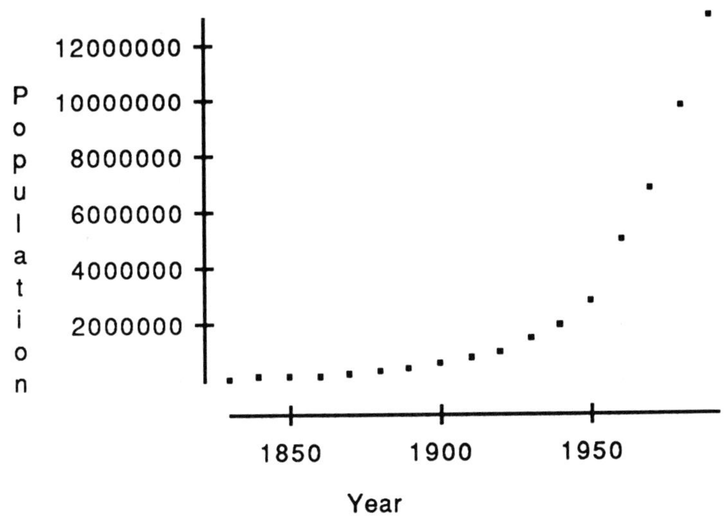


The relationship between population and year is dramatically non-linear. We might think about fitting a quadratic model to these data. However, a transformation is helpful here. A commonly used transformation is the logarithm (the base does not matter, but natural log - base e - is a typical choice). Taking the natural logarithm of Y changes the scatter plot to this:

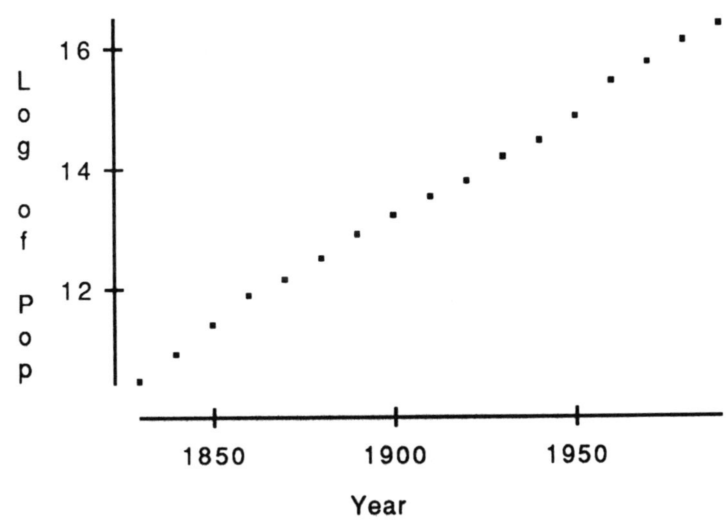


[1]Based on data from the U.S. Bureau of the Census.

**EXAMPLE 2**

Transformations can also be used to stabilize the variance of the response variable in a regression problem. In Example 3 of Section 5.1, we saw increasing variability in library science materials budget, Y, as number of science faculty, X, increased. Taking the square root of each value of Y changes the scatter plot:

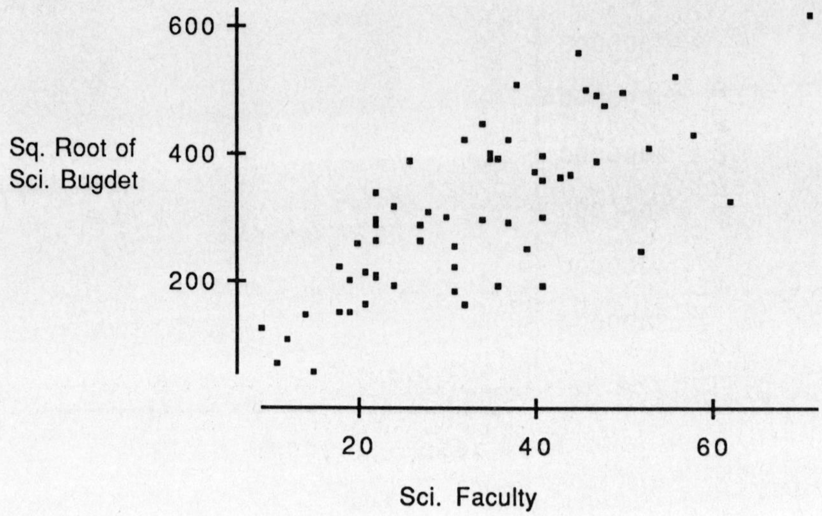


The residual plot from fitting a straight line to these data shows that the variability in the transformed Y values is fairly constant.

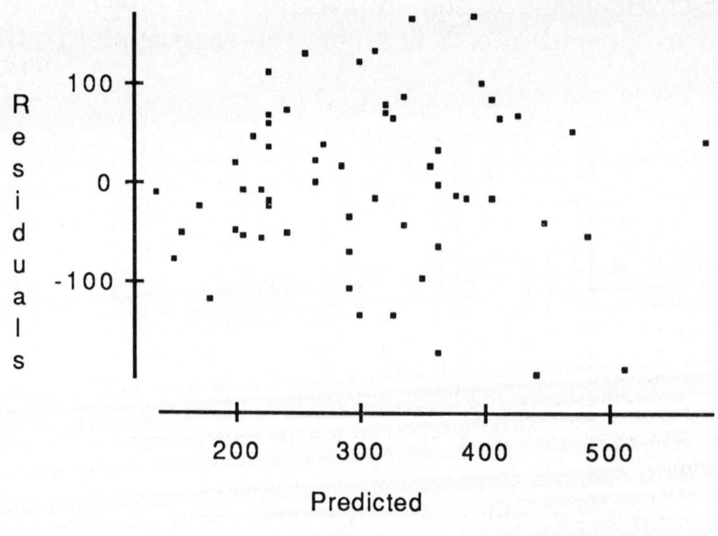


This is an improvement over the residual plot for the untransformed data (shown in Section 5.1).

Although the logarithm and the square root are the most commonly used transformations, many others are possible, including the reciprocal (1/Y) and the reciprocal square root ($1/\sqrt{Y}$). We might also consider transforming X. Generally there is no single "correct" transformation; we may try several, choosing the one that best helps us understand the relationship between the variables.

**EXAMPLE 3**

In some situations we need to transform both Y and X. The following scatter plot shows the relationship between brain weight, in grams, and body weight, in kilograms, for 39 species of animals[2].

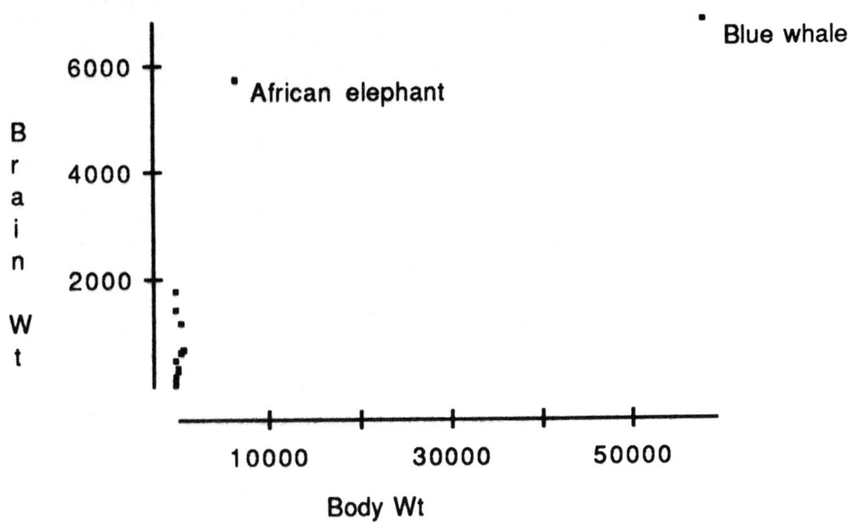


Most of the points are clustered together in the lower-left corner, so that all we learn from this plot is that the blue whale and the African elephant are unusually large. However, if we take the natural logarithm of each variable we see a linear relationship.

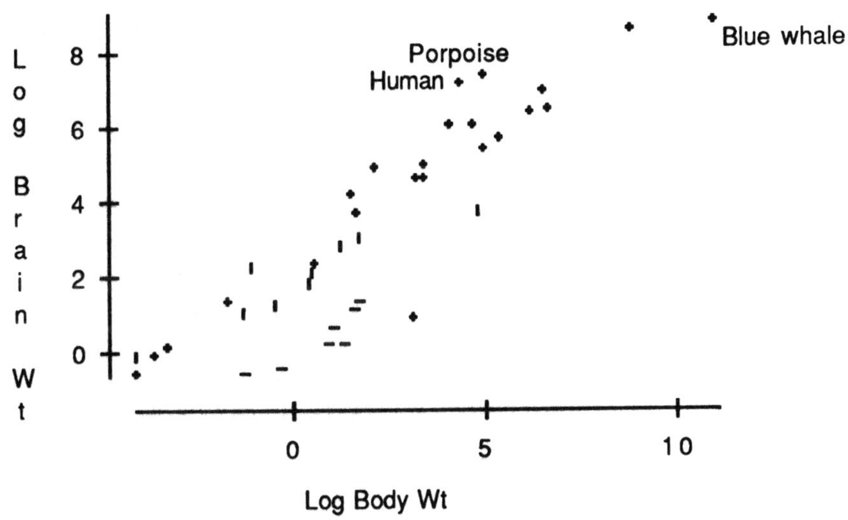


Here mammals are shown with the plotting symbol +, birds with the plotting symbol I, and fish with the plotting symbol -.

---

2 These data are from G. Crile and D.P. Quiring (1940), "A Record of the Body Weight and Certain Organ and Gland Weights of 3690 Animals," *Ohio J. of Science*, **5**:219-259. The raw data appear in the Appendix.

**Example 1**

The following scatter plot shows the relationship between the number of burglaries in 1988, Y, and the population, X, of large cities in Massachusetts[1], along with the regression line

$$\hat{Y} = -7.9 + 0.01656X.$$

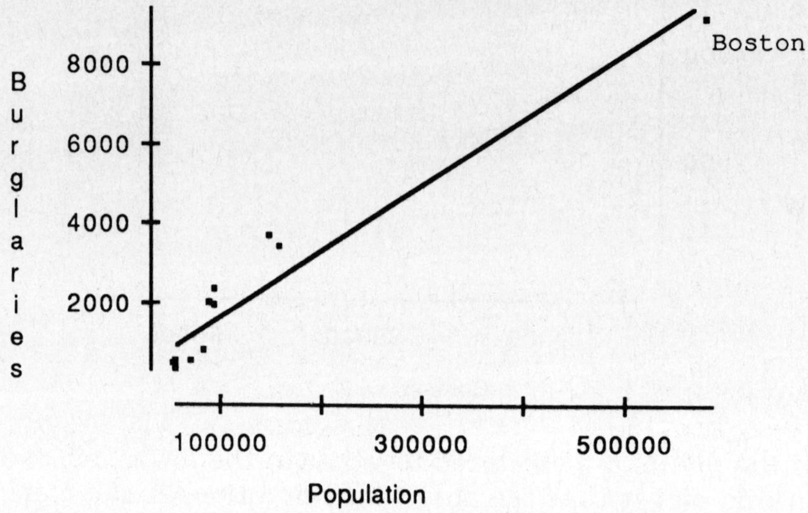


Note that Boston stands out, both in population and in number of burglaries. When we fit the regression model, Boston is an *influential point* because it has an exceptionally large influence on the fitted model.

If we delete Boston from the data set, the fitted model changes to $\hat{Y} = -1371.6 + 0.03248X$, which is shown on the following plot. The slope is now nearly double what it was with Boston in the data set. Deleting any other city from the original data set has essentially no effect on the fitted model.

---

[1]These data are taken from the 1988 *Uniform Crime Reports* of the F.B.I., U.S. Dept. of Justice, Washington, D.C. Here "large" cities are those with populations in excess of 50,000.

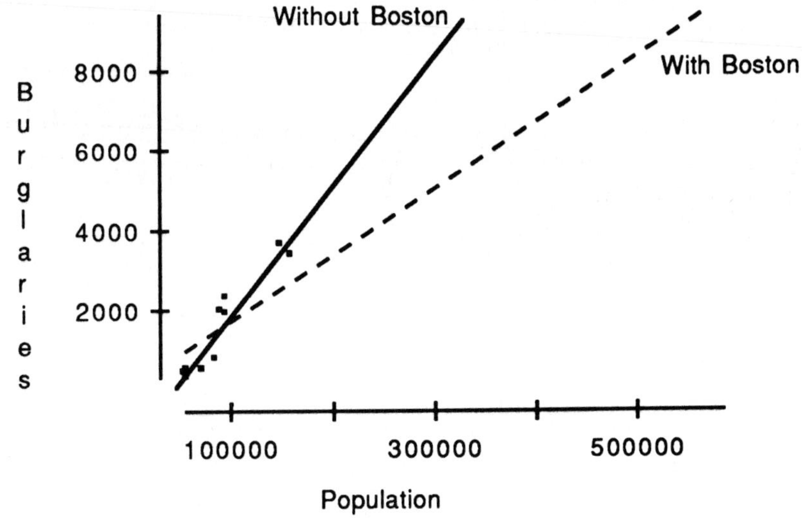


Although we can identify outliers on a residual plot, it is not always easy to spot the points that have large influence on the fitted regression model. The diagnostic statistic *Cook's distance* measures the amount of influence that each point has. Many computer packages calculate Cook's distance for each point when a regression model is fit. Here is a histogram of Cook's distance for the burglary data. Note that there is one extreme point here (Boston).

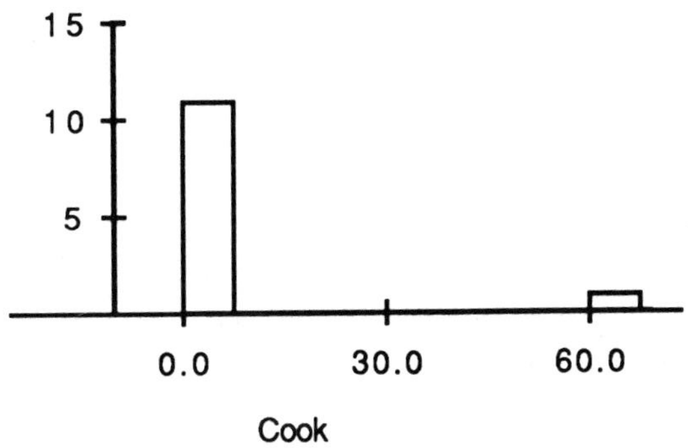


**Example 2**

The median age of residents varies from state to state, as does the percentage of persons over age 65 in the state. The following scatter plot shows the relationship between these variables.

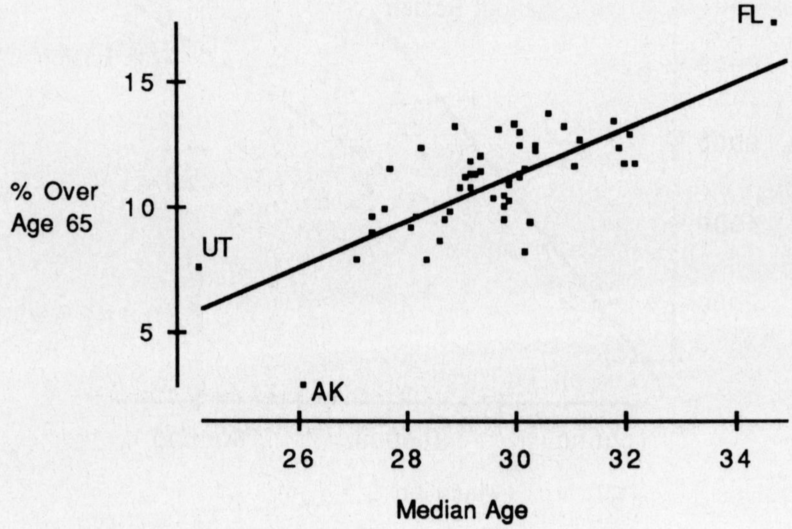


There are three unusual points here - Florida, Utah, and Alaska. The histogram of Cook's distance shows that Florida and Utah have somewhat more influence on the regression line than do most states. However, they do not deviate from the pattern created by the other states.

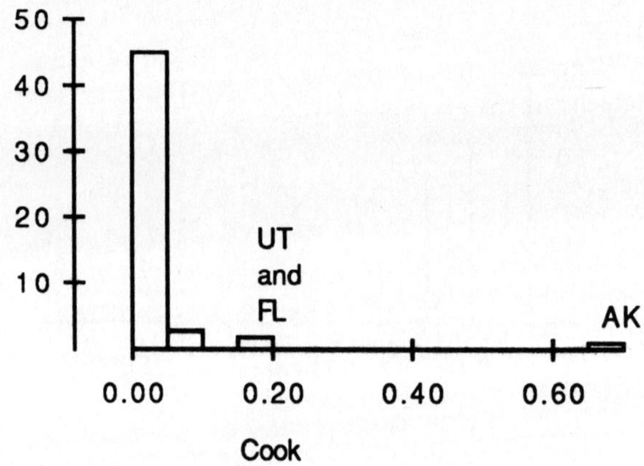


Alaska, on the other hand, not only has a great deal of influence in the fitted regression model, but is also an outlier, as can be seen clearly on the residual plot.

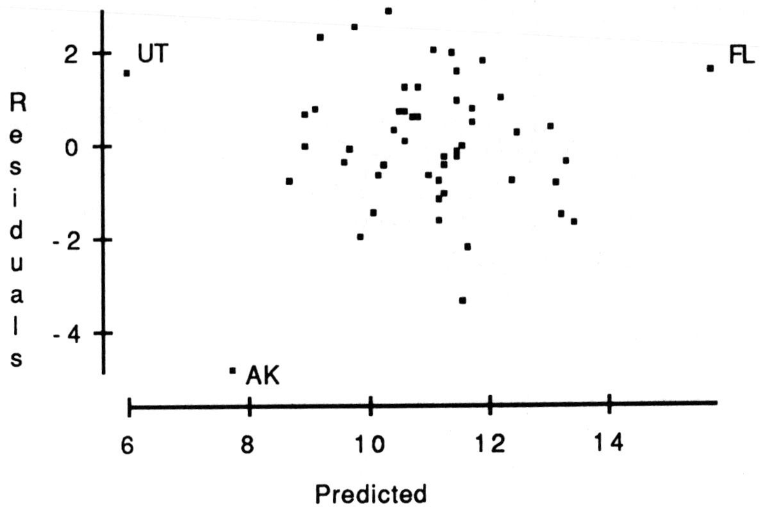
Residuals
2
0
-2
-4
UT
FL
AK
6
8
10
12
14
Predicted

6.1 Fitting Multiple Regression Models

**Example 1**

In Chapter 4 we used a regression model to study the relationship between the per capita income level of a state (*percap$*) and the percentage of adults age 25 years and older in the state who have completed at least four years of college (*college%*). The relationship between *percap$* and *high school%*, the percentage of adults age 25 years and older in the state who have completed four years of high school, is also quite strong (r=.621).

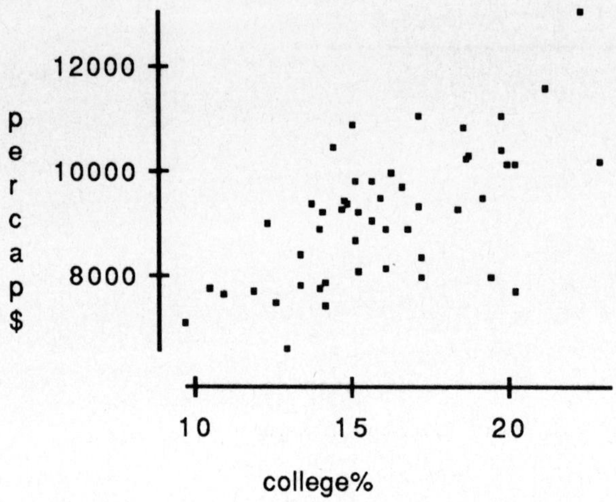


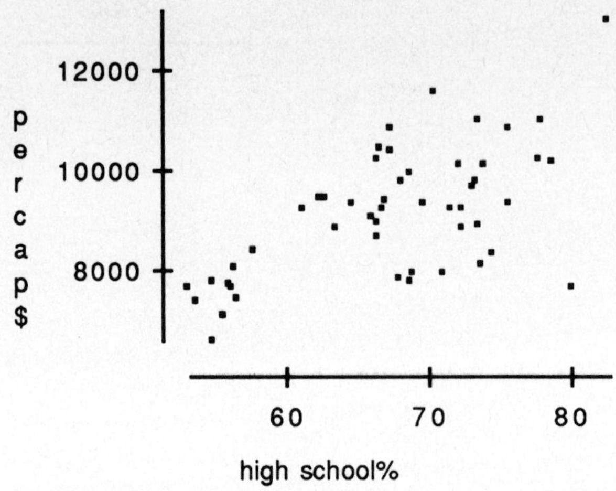


We can see increasing trends in both of the scatter plots above. If we fit a regression model in each case, we obtain the following:

$$\widehat{percap\$} = 4696 + 276*college\%$$

$$\widehat{percap\$} = 2015 + 106*high\ school\%$$

We can use *college%* and *high school%* <u>together</u> by fitting a *multiple regression model* of the form

$$Y = \beta_0 + \beta_1 X_1 + \beta_2 X_2 + \varepsilon$$

where $Y = percap\$$, $X_1 = college\%$, and $X_2 = high\ school\%$. The mathematics involved in fitting such an equation using least-squares is a bit messy, but many statistical computer packages will do the appropriate calculations. The fitted model is

$$\widehat{percap\$} = 2624 + 177*college\% + 54*high\ school\%$$

This equation indicates that as *college%* goes up, so does *percap$*. Likewise, as *high school%*, so does *percap$*. The interpretations of the coefficients "177" and "54" are a bit tricky, however. Our model says that both *college%* and *high* *school%* affect *percap$*. Thus, when we talk about the effect of *college%* we cannot ignore the fact that *high school%* is in the equation. Rather, we would say "*college%* has a positive association with *percap$*, controlling for *high school%*," which is to say that the effect on *percap$* of a large value of *college%*, beyond the effect of *high school%*, is positive. This is the benefit of fitting a multiple regression model: we can assess the effect of a variable after adjusting for the effects of other variables.

As a check on the appropriateness of the model, we plot the residuals from the model against the predicted values:

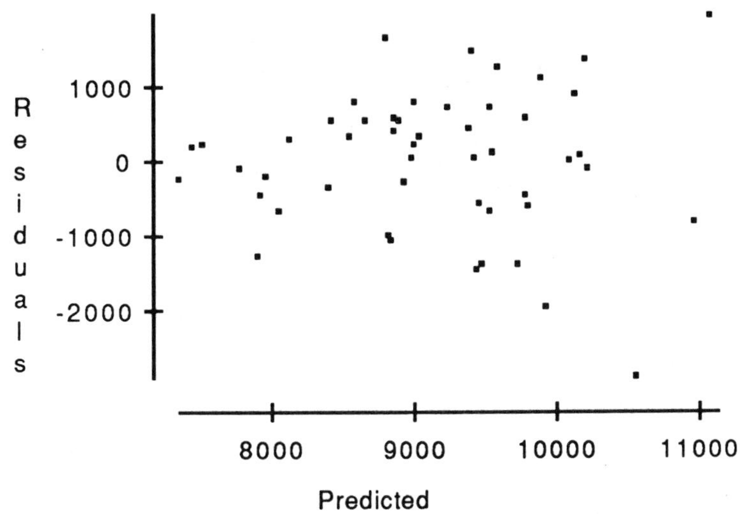


This residual plot does not reveal any serious problems or outliers (although there is a hint of non-constant variance).

In fitting multiple regression models, we do not have to restrict ourselves to using only two independent variables. (We can include $X_3$, $X_4$, etc.) But we should always examine scatter plots and residual plots, so that we do not miss any curves or outliers in the data set.

**Example 2**

Consider the relationship between Y, reading score, and $X_1$, height, for a group of children. (These data are hypothetical.)

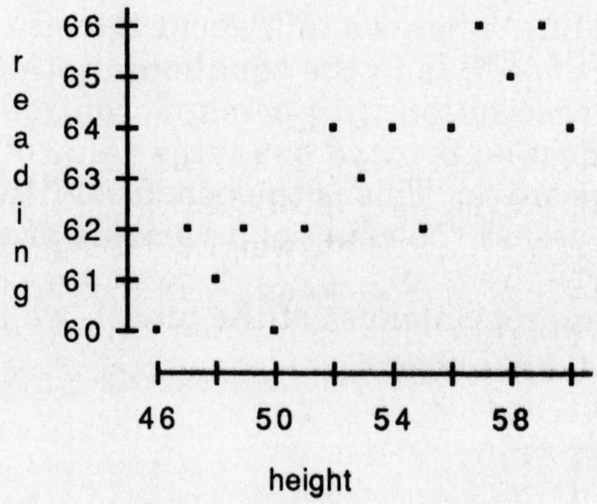


The fitted equation here is

$$\widehat{reading} = 44 + .357*height.$$

There are three age groups in this data set, each of which forms an M shape in the scatter plot. In this case, when we control for the variable $X_2$ = age we get

$$\widehat{reading} = 59 + 0*height + 2*age.$$

As we would expect, controlling for age gives quite a different picture of things than does ignoring age. We would interpret the last equation as "Controlling for age, there is no association between height and reading score."

When we fit a multiple regression model one of the values produced by the computer is the multiple coefficient of determination, denoted by $R^2$. This value measures how well the independent variables (the Xs), taken as a group, do in helping us estimate (or predict) Y. If $R^2$=1 then the regression equation fits the data perfectly. If $R^2$=0 then knowing the values of $X_1$, $X_2$, etc. is of no help in estimating Y. Usually, we are somewhere between these two extremes. For instance, in Example 1, $R^2$=.466, or 46.6%. This means that 46.6% of the variation in *percap$* is "explained" by the two variables *college%* and *high school%*. The word "explained" is in quotes because it can be dangerous to imply a cause-and-effect relationship between the variables.

A special case of multiple regression is known by the name of *analysis of covariance*, or ANACOVA. In his book *The Mismeasure of Man*, Stephen Jay Gould discusses the theory put forth by Paul Broca and his followers in the nineteenth century that intelligence depends upon brain size. (This theory was used to support the claim that women are less intelligent than men - they have smaller brains!) Broca's theory was challenged by Louis Pierre Gratiolet, using data that showed that German brains were 100 grams heavier, on average, than French brains. One of Broca's colleagues, M. de Jouvencel, responded by saying that if one were to make a "correction" for body size (height and weight) the apparent German advantage would disappear. The idea underlying de Jouvencel's argument is that we should control for body size, rather than ignore it. (Gould wonders why Broca and his followers avoided doing this when comparing the brain sizes of men and women.)

Within the framework of modern multiple regression we would control for body size by fitting an equation of the form

$$Y = \beta_0 + \beta_1 X_1 + \beta_2 X_2 + \varepsilon$$

where Y = brain size, $X_1 = 1$ for German, 0 for French, and $X_2$ = body size. (For the variable "body size" we might use height, or we might replace this single variable with a group of variables such as height, weight, muscular mass, etc., and have a longer regression equation.) In this case our primary interest is in comparing two groups, namely the German and French peoples, and the other factor, body size, is in the equation so that we can remove its confounding effect (control for it). We call body size a covariate.

In this regression model the variable $X_1$ is called a *dummy variable*, since we create it by arbitrarily assigning the value 1 to the Germans and the value 0 to the French. Thus, this model effectively provides two regression lines with a common slope:

$$Y = \beta_0 + \beta_1 + \beta_2 X_2 + \varepsilon \qquad \text{for the Germans}$$

and

$$Y = \beta_0 + \beta_2 X_2 + \varepsilon \qquad \text{for the French.}$$

The difference between the two nationalities is quantified in the model by the additive constant $\beta_1$.

**Example 3**

Table 6.1 lists various nutrition characteristics of a one-ounce (28.35 gram) serving of each of 36 breakfast cereals[1]. To study the relationship between calories, sugar, and fiber we can plot sugar vs. fiber

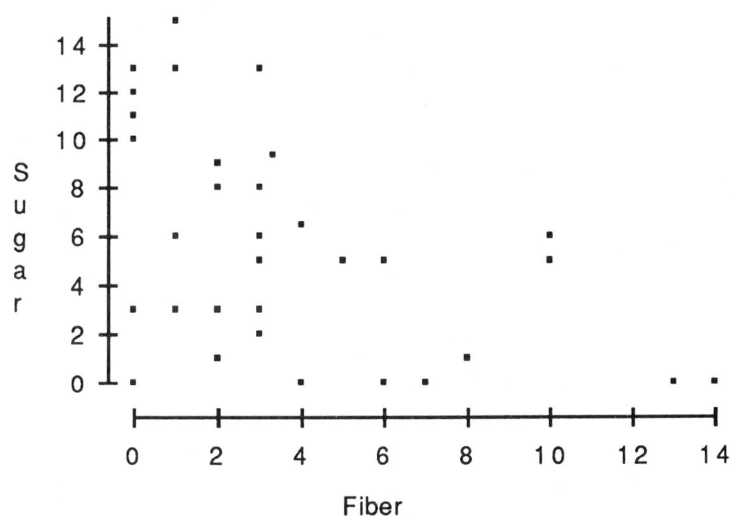


[1] These data were collected by Patricia Benedict, Ronald Brahler, and Kenneth Motz, who read the nutritional labels on the boxes, in an attempt to learn whether cereals high in fiber are also high in sugar, sodium, and calories. The sodium data are not presented here. The cereals are all of those that were sold at Russo' Stop & Shop in University Heights, OH, in July, 1990.

calories vs. fiber

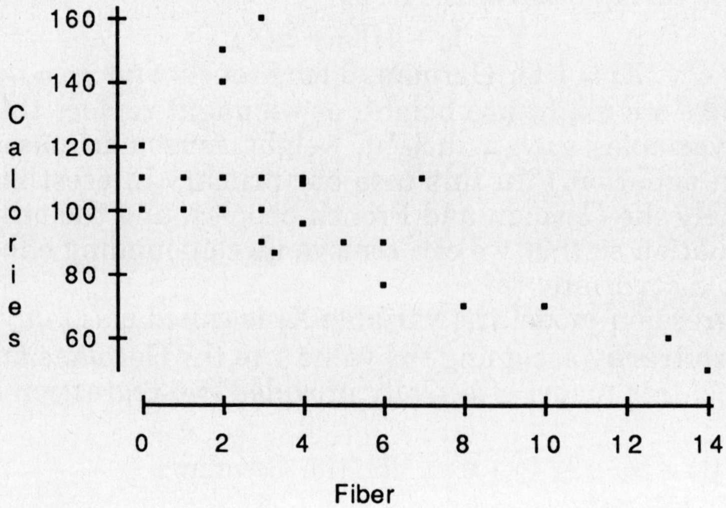


and calories vs. sugar

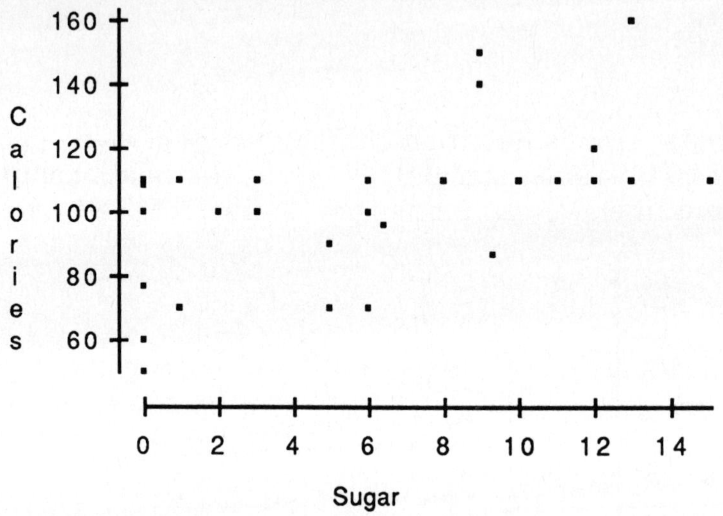


Putting these scatterplots together we have a *scatterplot matrix*. This graph allows us to look easily at each pair of variables. Notice that as calorie level goes up sugar content goes up but fiber level goes down.

**Table 6.1**

| Cereal | Calories | Sugar (g) | Fiber (g) |
|---|---|---|---|
| Common Sense Oat Bran | 100 | 6 | 3 |
| Product 19 | 100 | 3 | 1 |
| All-Bran with extra fiber | 50 | 0 | 14 |
| Just Right | 140 | 9 | 2 |
| Original Oat Bran | 70 | 5 | 10 |
| Heartwise | 90 | 5 | 6 |
| Special K | 110 | 3 | 0 |
| Oatbake Raisin Nut | 110 | 8 | 2 |
| Kenmei Rice Bran | 150 | 9 | 2 |
| Nutri Grain | 100 | 2 | 3 |
| Shredded Wheat Squares | 90 | 5 | 3 |
| Oatmeal Raisin Crisp | 110 | 6 | 1 |
| Whole Wheat Total | 100 | 3 | 3 |
| Cheerios | 110 | 1 | 2 |
| Total Raisin Bran | 86.7 | 9.3 | 3.3 |
| Wheaties | 100 | 3 | 3 |
| Raisin Nut Bran | 110 | 8 | 3 |
| Wheat Chex | 100 | 3 | 2 |
| Batman | 110 | 10 | 0 |
| Ninja Turtles | 110 | 13 | 0 |
| Cap'n Crunch | 120 | 12 | 0 |
| Trix | 110 | 12 | 0 |
| Frosted Flakes | 110 | 11 | 0 |
| Honey Smacks | 110 | 15 | 1 |
| Froot Loops | 110 | 13 | 1 |
| Puffed Rice | 100 | 0 | 0 |
| Mueslix Crispy Blend | 160 | 13 | 3 |
| Uncle Sam | 110 | 0 | 7 |
| 100% Bran | 70 | 6 | 10 |
| Fruit & Fiber | 96 | 6.4 | 4 |
| Bran Flakes | 90 | 5 | 5 |
| Bran Buds | 70 | 1 | 8 |
| Fruit'n Oat Bran | 110 | 0 | 4 |
| Fruit'n Oat Bran Crunch | 60 | 0 | 13 |
| Hodgson's Mill Wheat | 108 | 0 | 4 |
| Hodgson's Mill Oat Bran | 77 | 0 | 6 |

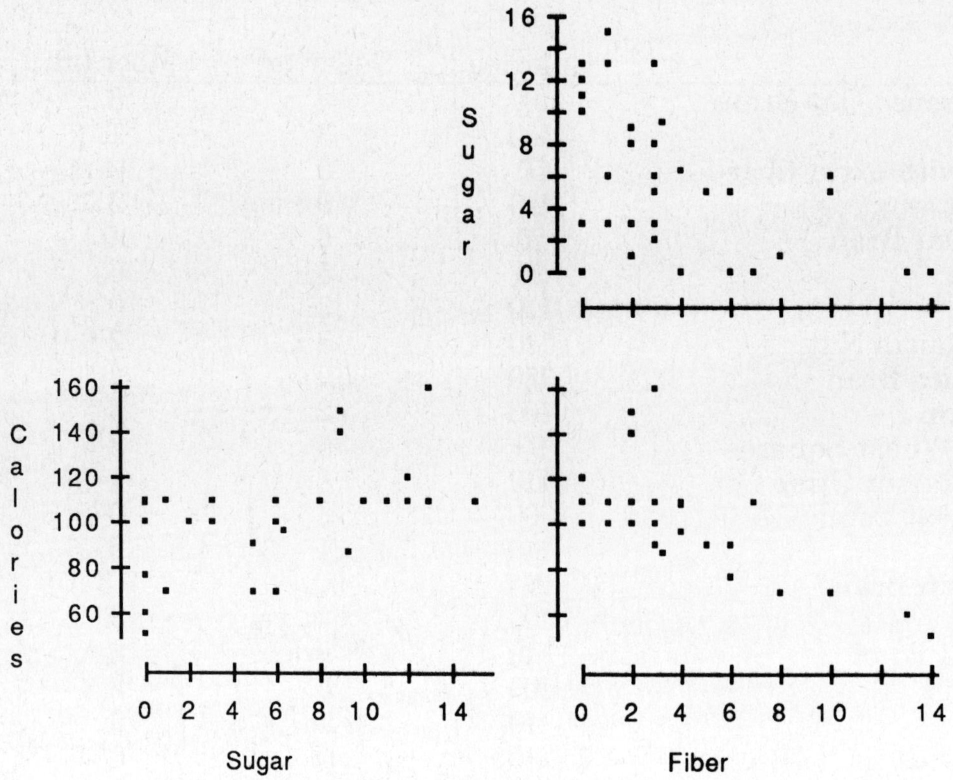


The relationship between Calories and Sugar is moderately strong: r = .515. However, Fiber is more strongly correlated with Calories: r = -.715. When we regress Calories on Fiber we get

$$\widehat{\text{Calories}} = 117.4 - 4.4\text{*Fiber}$$

and $R^2$ = 51.1%. Adding Sugar to the model changes the equation to

$$\widehat{\text{Calories}} = 109.3 + 1\text{*Sugar} - 3.7\text{*Fiber}$$

but only increases $R^2$ to 54.4%. Thus, controlling for Fiber being in the model, the relationship between Sugar and Calories is fairly weak.

The residual plot for this multiple regression has an unusual feature: Since the data are rounded and tend to fall into discrete groups (e.g., there are 13 cereals with a calorie count of 110), there are sets of points in the residual plot that fall more or less on a line. As the predicted values, $\hat{Y}$, go up, the residuals in these groups go down.

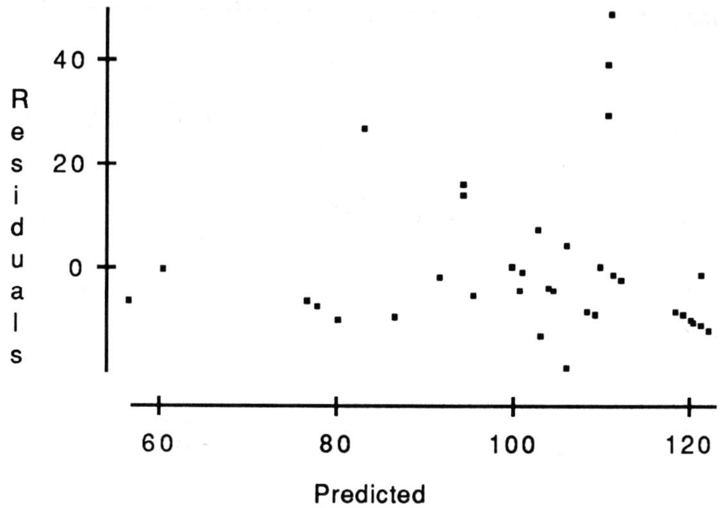


<u>Exercise</u>

1. There is a group of 7 cereals in Table 6.1, starting with Batman and ending with Froot Loops, that are marketed primarily for children. Delete these cereals from the data set and perform a new analysis. How does deleting this group change matters?

6.2 Interactions

**Example 1 - Highway Fatality Rate**

Table 6.2 presents rural interstate highway fatality rates in Ohio for the years 1967 - 1985[1].

**Table 6.2**

| Year | Rural Interstate Fatality Rate | Year | Rural Interstate Fatality Rate |
|---|---|---|---|
| 1967 | 2.36 | 1977 | 0.91 |
| 1968 | 2.36 | 1978 | 0.79 |
| 1969 | 2.64 | 1979 | 0.65 |
| 1970 | 2.06 | 1980 | 0.58 |
| 1971 | 2.10 | 1981 | 0.69 |
| 1972 | 1.67 | 1982 | 0.53 |
| 1973 | 1.54 | 1983 | 0.75 |
| 1974 | 0.78 | 1984 | 0.57 |
| 1975 | 0.85 | 1985 | 0.63 |
| 1976 | 0.50 | | |

The scatter plot of rate against year shows a downward trend over time. There are two groups of points in the data set: years up to and including 1973, when the speed limit was 70 mph, are plotted with the symbol ·; years after 1973, when the speed limit was 55 mph, are plotted with the symbol **x**.

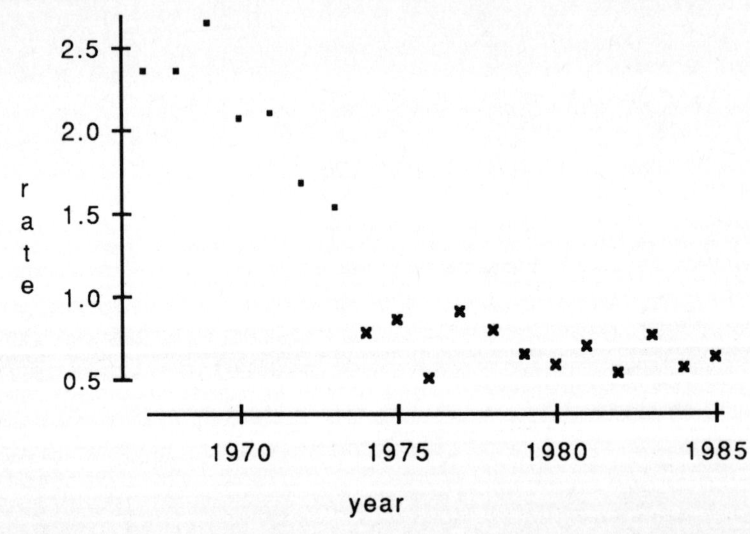


The overall trend is not linear, but within each group the relationship between rate and year is linear. We can model changes over time in the fatality rate using the regression equation

(*) $$Y = \beta_0 + \beta_1 X_1 + \beta_2 X_2 + \beta_3 X_1 X_2 + \varepsilon$$

[1] These data were collected by the Federal Highway Administration. The rural interstate highway fatality rate is the number of fatalities per 100 million miles traveled.

where Y = fatality rate, $X_1$ = year, and $X_2$ is a dummy variable that takes on the value 0 for the years 1967-73 and 1 for the years 1974-85. The term $\beta_3X_1X_2$ is an *interaction* term. To see the effect of this term, consider the two possible values for $X_2$.

(a) When $X_2 = 0$ (for 1967-73), the regression model is

$$Y = \beta_0 + \beta_1X_1 + \varepsilon$$

(b) When $X_2 = 1$ (for 1974-85), the regression model is

$$Y = \beta_0 + \beta_1X_1 + \beta_2 + \beta_3X_1 + \varepsilon$$

which take be written as

$$Y = (\beta_0 + \beta_2) + (\beta_1 + \beta_3)X_1 + \varepsilon$$

Thus, we get two lines, one for each group of years. The fitted equation is

$$\hat{Y} = 310.269 - 0.15643\ X_1 - 277.122\ X_2 + 0.14003\ X_1X_2$$

This interaction model says that year ($X_1$) has an effect on the fatality rate and speed limit ($X_2$) has an effect on fatality rate, but these effects are not separate from each other. Rather, year and speed limit *interact* in affecting rate. There is a negative slope between year and rate, but this slope depends on $X_2$. When $X_2 = 0$ the slope is -0.156429; when $X_2 = 1$ the slope is -0.016399. Thus, when $X_2 = 0$ the effect of year on rate is large (the line is steep), but when $X_2 = 1$ the effect is small. The two lines are shown on the following scatter plot[2].

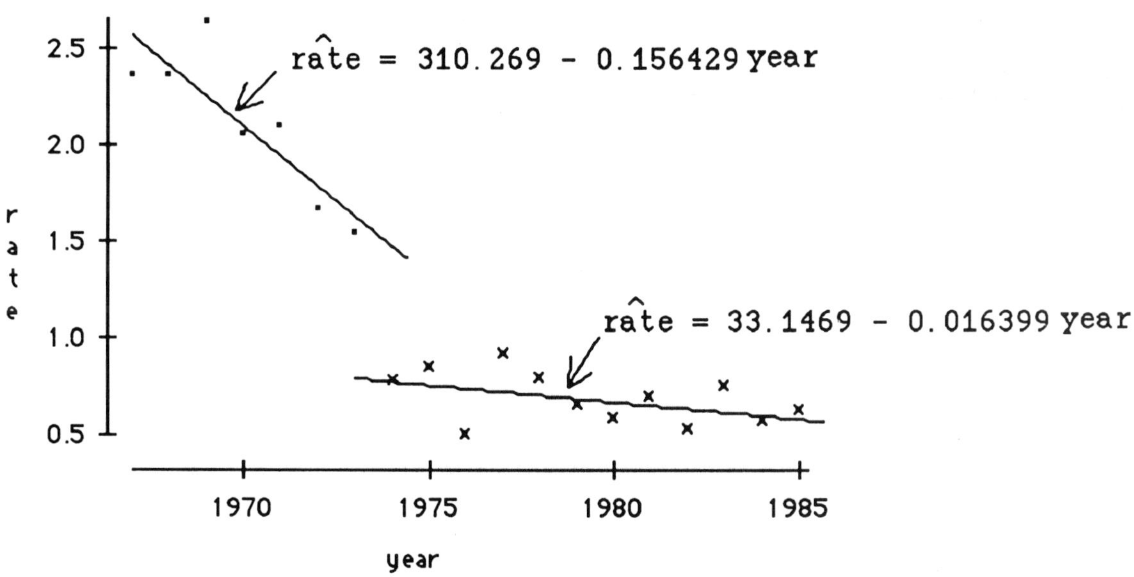


[2] We could divide the data set into two parts and fit separate regression lines to each, but if we believe that the variance of the error, ε, is the same in each group then the regression model (*) is appropriate. Moreover, it is convenient to work with a single regression model, rather than fitting two equations.

The plot of residuals against predicted values has two clusters of points. If we plot the residuals against year, we see that the variability is fairly stable.

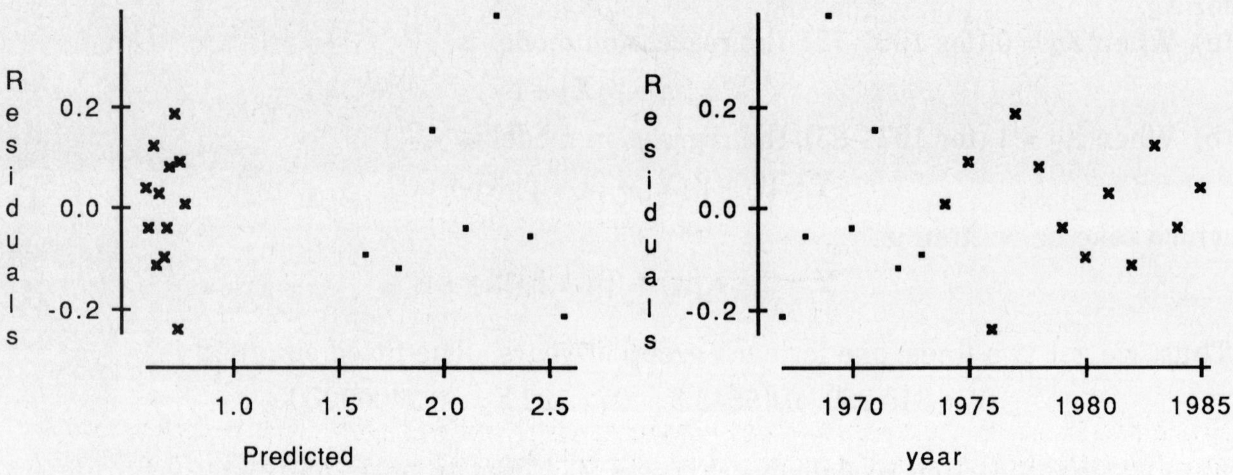


**EXAMPLE 2 - Response Time**

Bill Friedman, a professor of psychology at Oberlin College, measured the length of time, Y, that it took a young child to correctly report the time when reading a digital clock and when reading an analog clock. He repeated this for several children under differing settings[3]. Table 6.3 shows the average response time for each of four combinations of clock type and stimulus time.

**Table 6.3**

| | | Stimulus Time | |
|---|---|---|---|
| | | 4:00 | 4:43 |
| Clock Type | Analog | 2.10 | 9.92 |
| | Digital | 1.28 | 1.94 |

We can plot the average response time at each combination of clock type and stimulus time in an *interaction plot*.

[3] These data are from part of a larger experiment, a related aspect of which is reported in W.J. Friedman and F. Laycock (1989), "Children's Analog and Digital Clock Knowledge," *Child Development*, **60**:357-371.

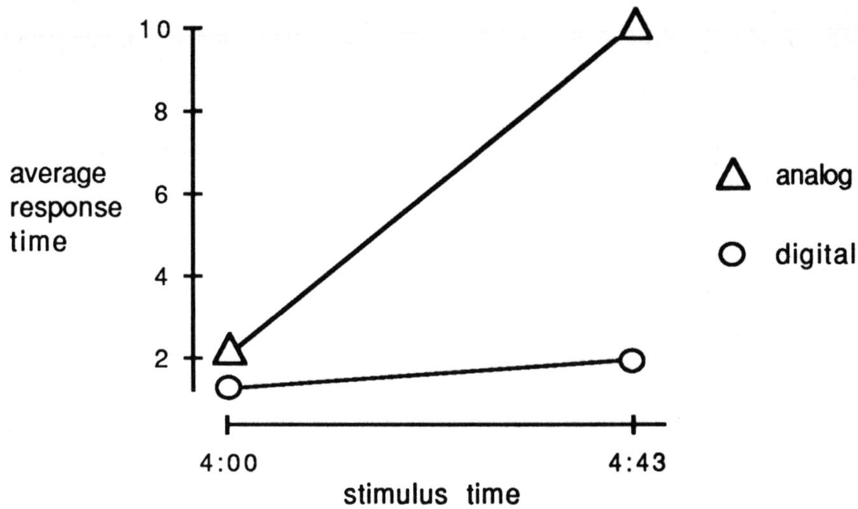


Since the lines connecting average response time within each clock type are not parallel, the plot shows that there is an interaction between clock type and stimulus time in their effect on response time. Stimulus time has little effect on response time for a digital clock, but it has a large effect on response time for an analog clock.

Let $X_1 = 0$ for analog and 1 for digital, and let $X_2 = 0$ when stimulus time is 4:00 and 1 when stimulus time is 4:43. We call $X_1$ and $X_2$ "dummy variables." The fitted interaction model is

$$\hat{Y} = 1.28 + 0.66\,X_1 + 0.82\,X_2 + 7.16\,X_1X_2$$

(Note: this model, in which all of the X variables are dummy variables, is an analysis of variance (ANOVA) model.)

**Example 3 - Laughter and Funniness**

Do people always laugh at something they find funny? Porterfield, Mayer, et al.[4], conducted an experiment in which subjects listened to tapes of several "bloopers" and, independently of each other, rated each blooper on a funniness scale. The experimenters recorded the level of audible laughter of each subject for each blooper and then averaged the funniness ratings and laughter levels for each subject.

Subjects were measured on a scale of private self-consciousness (PSC). Among those subjects with low PSC ratings (i.e., those who are outwardly directed), the relationship between laughter and funniness is weak. However, some subjects heard a laugh-track with the bloopers, but others did not, and the presence or absence of a laugh-track has an effect here: the correlation between laughter and funniness is positive ($r = .378$) if the bloopers are played with a laugh-track, but it is negative ($r = -.312$) if the bloopers are played without a laugh-track. The following scatter plot shows these two sets of points using different

[4] For a full report of this experiment, see A.L. Porterfield, F.S. Mayer, et al. (1988) "Private Self-consciousness, Canned Laughter, and Responses to Humorous Stimuli," *J. of Research in Personality* **22**:409-423.

plotting symbols: **x** if there was a laugh-track, · if there was not. Each point represents one person.

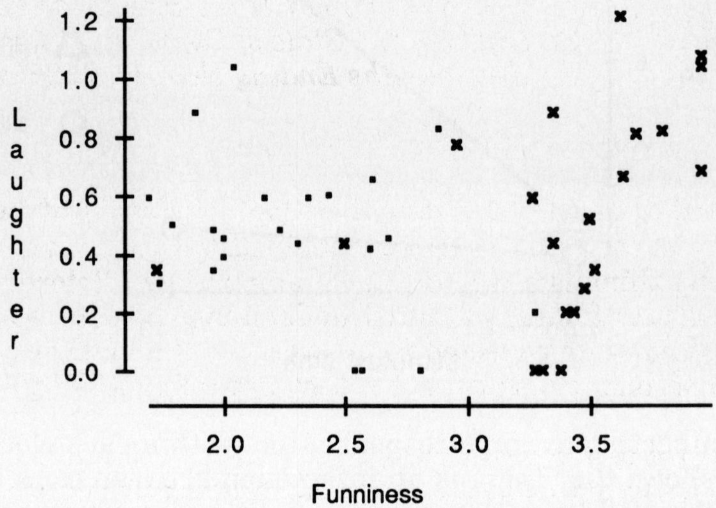


Note that most of the subjects who gave high funniness ratings heard a laugh-track with the bloopers, and that the higher the funniness ratings they gave, the more they laughed.

A fitted regression model for these data is

$$\hat{Y} = 0.92 - 0.2\,X_1 - 1.29\,X_2 + 0.47\,X_1X_2$$

where Y = laughter level, $X_1$ = funniness rating, and $X_2$ is a dummy variable that equals 1 if there is a laugh-track and 0 otherwise. The coefficient of the interaction term $X_1X_2$ is positive, reflecting the fact that when there is a laugh-track (i.e., when $X_2 = 1$) the slope between funniness, $X_1$, and laughter, Y, is greater than it is when there is no laugh-track. Learning that funniness and the presence or absence of a laugh-track interact in their effect on laughter (for persons who have low PSC ratings) was an important finding in this study.

"Nothing is easier than to get some sort of figures.... Nothing is harder to get than figures on which we can rely."

*First Annual Report of the Bureau of Labor Statistics of the State of Connecticut for the Five Months Ending Nov. 30, 1885,* p.3.

## 7.1 Surveys

As suggested by the above quote, the quality of data is of paramount importance in any study. Conducting a sophisticated analysis of data that measure the wrong thing is a waste of time and energy. Before you can draw conclusions from a set of data, you must understand the source of the data and the question the data were collected to answer. As the following example shows, even a mountain of data are worthless if they are collected in a poor fashion.

**Example 1 - The *Literary Digest* Poll**

In 1936 the *Literary Digest*, a magazine that conducted the best known polls at that time, mailed questionnaires to 10 million persons, asking them how they intended to vote in the upcoming presidential election. Based on 2.4 million returned questionnaires, the *Digest* predicted that Kansas governor Alf Landon would win over Franklin D. Roosevelt by a landslide margin of 57% to 43%. Most political observers thought that Roosevelt would win, and he did, 62% to 38%. Thus, the *Literary Digest* was off by 19 percentage points.

What went wrong with the poll? There were two major problems. First, the *Digest* did not send questionnaires to a random sample of potential voters; rather, they drew their sample from such sources as telephone books and lists of automobile owners. This sampling scheme under-represented the poor and introduced a *selection bias* in favor of the Republican candidate, Landon.

The second problem was that only 24% of those who received questionnaires mailed them back; the other 76% did not respond. Those who responded painted one picture of the electorate, but *nonresponse bias* made this picture quite inaccurate[1]. For example, the *Digest* mailed questionnaires to one-third of the registered voters in Chicago, but only 20% responded. Over half of the respondents favored Landon, but in the election Roosevelt carried Chicago by a wide margin.

The *Literary Digest* prediction was based on a staggering number of returned questionnaires, but these were not from a random sample of the population. Having a large sample size does not correct the problems of selection bias and nonresponse bias.

**Example 2 - The National Newspaper Poll**

During Labor Day week in 1985 several newspapers across the U.S. conducted the National Newspaper Poll by inviting readers to answer a 172-item questionnaire covering a wide variety of topics. In addition to problems of

---

[1] In his paper "The Literary Digest Poll: Making of a Statistical Myth" (*American Statistician*, 1976, **30**:184-5), M. C. Bryson argues that nonresponse bias, rather than selection bias, was the leading cause of the error in the *Literary Digest* prediction.

selection and nonresponse bias, reports of the findings of this poll featured another potential problem with a survey.

An article analyzing the responses carried the headline "Morality has slipped, poll finds[2]." However, none of the 172 items on the questionnaire asked readers about their own morality. Rather, the questionnaire included a series of items asking readers what percentage of Americans they would guess obey each of the Bible's Ten Commandments, and what percentage they guess obeyed the Commandments 10 years earlier. The pollsters found that the typical respondent thought that compliance with the Commandments had dropped over time; hence the headline. Another headline read "Poll finds youth are pessimistic[3]," but is based on responses from adults, not from youth. The point is that one must be careful to ascertain what question was asked, and of whom, before one can interpret the findings of a poll.

**Example 3 - Fidelity**

In 1987 Shere Hite reported the results of a survey she had made of 4,500 women[4]. Among her findings was that 70% of the women married five years or more said they had had an extramarital affair. She gathered her data by mailing questionnaires to 100,000 members of various women's groups.

At roughly the same time, advice columnist Abigail Van Buren (Dear Abby) asked her readers "Have you ever cheated on your mate?" She heard from 150,000 women, only 15% of whom said they had cheated[5].

Whom should we believe? Since both of these surveys have massive selection and nonresponse biases, neither can be trusted. The actual percentage of married women who have had affairs is probably somewhere between 15% and 70%, but to get people to tell you the truth about such an issue probably requires the randomized response method - see Section 7.2.

Given all of the things that can go wrong with a survey (the wrong question may be asked, there may be sampling bias or nonresponse bias, the race of the pollster may influence responses[6], etc.), how can one be sure of the results? Here is a list of things to check when reading about a survey.

1) Is the population clearly defined? Is it the proper population to study? Is the population changing over time?
2) Exactly what question was asked? Who asked it?

---

[2] This article, published in the *Gainesville Sun*, 10 November 1985, p. 6B, was one of a series of articles on the National Newspaper Poll by A. S. Gillman, D. C. MacKay, and P. Sullivan.

[3] A. S. Gillman, D. C. MacKay, and P. Sullivan, *Gainesville Sun*, 12 November 1985, p. 1.

[4] *Time*, 12 October 1987, p. 68-73.

[5] *The Cleveland Plain Dealer*, 21 February 1988, p. 1G.

[6] It seems that prior to the 1989 gubernatorial election in Virginia, white voters were reluctant to admit to black pollsters that there were going to vote for the white candidate, Marshall Coleman, rather than the black candidate, Douglas Wilder. See *Time*, 20 November 1989, p. 56.

3) How was the sample drawn? Was it a random sample? (That is, did each member of the population have the same chance of being sampled?[7]) Or was it a sample of convenience?
4) What was the response rate? If not 100%, is there reason to suspect nonresponse bias?
5) What was the sample size? Is a margin of error reported?

---

[7] Technically, a random sample of size n is one in which each possible subset of the population of size n has the same chance of being selected.

## 7.2 Randomized response sampling

When pollsters ask "Do you think the President is doing a good job?" they assume that people will tell them the truth. But suppose the Internal Revenue Service asked a random sample of people "Did you cheat on your taxes this year?" Many, if not all, of those who cheated would lie about it. The same problem faces AIDS researchers who study the spread of the disease: There are some sensitive questions to which people do not respond truthfully.

In 1965 Warner[1] proposed a technique, called randomized response sampling, that is now widely used in asking sensitive questions. There are now several variations on this idea that have been proposed, but the following may be the most easy to use and to understand.

**Example 1 - Adultery**

What percentage of mathematics teachers have committed adultery? Let this unknown percentage be denoted by $\theta$. I used the randomized response method, with a convenience sample of 24 high school mathematics teachers[2], to estimate $\theta$.

First, I had each teacher, in private, toss two coins. Then I told them to secretly split themselves into two groups: Those who got heads on the first coin toss were to respond, with either "yes" or "no," to Item #1; those who got tails were to respond to Item #2. The two Items were:

Item #1: The second coin toss resulted in heads.
Item #2: I have committed adultery.

I explained that I had no way of knowing who, if any of them, would be responding to the sensitive question - had they committed adultery? - so they could freely tell the truth. I had them write their responses on slips of paper, which I collected.

A tally of the responses showed that 13 of the 24 had written "yes," so the sample percentage of yeses, denoted by $\hat{p}$, was 13/24 = 54.2% Although I had no way of knowing who, if any of them, had responded to Item #2, I estimated $\theta$ in the following way. I expect that, on average, half of them would get heads on the first coin toss and would respond to Item #1, and half would respond to Item #2. Furthermore, of those responding to Item #1, I expect half of them to get heads on the second coin toss and to write "yes." Thus, I expect 25% of them to end up at the top branch of the probability tree below. I expect half of them to respond to Item #2, so I expect $\theta$*50% of them to (truthfully) write "yes" by ending up at the third branch of the probability tree.

---

[1] S. L. Warner (1966), "Randomized Response: A Survey Technique for Eliminating Evasive Answer Bias," *J. Am. Statist. Assoc.* **60**:63-9.

[2] These data were collected at a Quantitative Literacy workshop for high school mathematics teachers, at John Carroll University, in July, 1990. The actual randomization method used was slightly different from the coin tossing described here, but is equivalent. Note that this was a sample of convenience and <u>not</u> a random sample. I chose this example to illustrate the idea of randomized response, not as an example of how to design a survey.

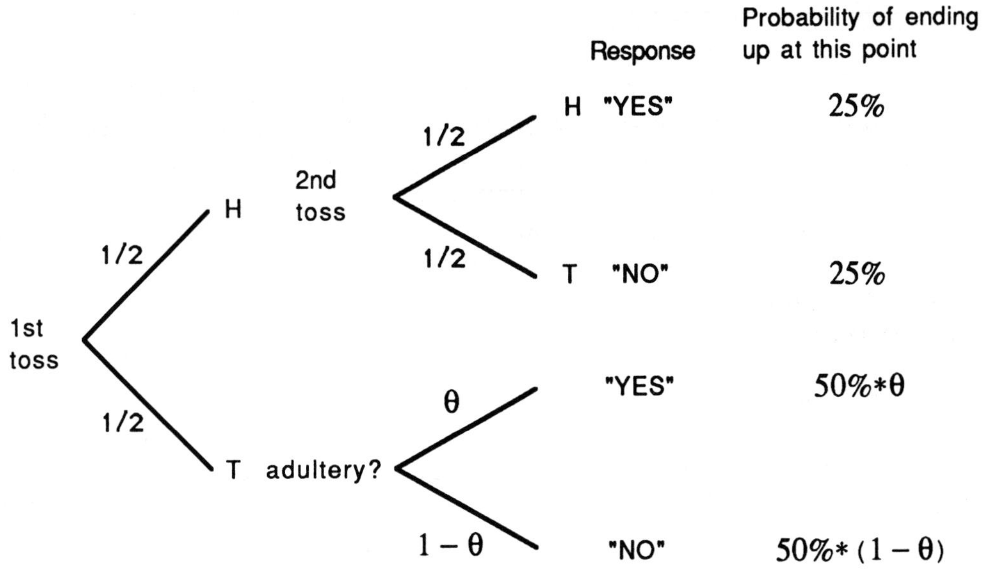


Overall, the expected value of the sample percentage of yeses is $.25 + .5*\theta$. Let us call this theoretical proportion p:

$$.25 + .5*\theta = p$$

We now use the observed sample percentage, $\hat{p}$, in place of the unknown p to get

$$.25 + .5*\theta = \hat{p}$$

Solving for θ, we have

$$\theta = 2*(\hat{p} - .25)$$

In this case,

$$\theta = 2*(.542 - .25) = .584$$

Thus, I estimate that 58% have committed adultery.

Another way to think about the probability tree is that we expect the 24 teachers to be split "12 and 12" on the first coin toss. Of the 12 we expect to end up in group #1, we expect 6 to get heads on the second toss and respond "yes". We don't know before-hand how many in group #2 we should expect to respond "yes", but since we had 13 yeses all together, we infer that 7 of them were from group #2. Thus, we end up thinking that 7 of 12 (58%) answered "yes" to Item #2.

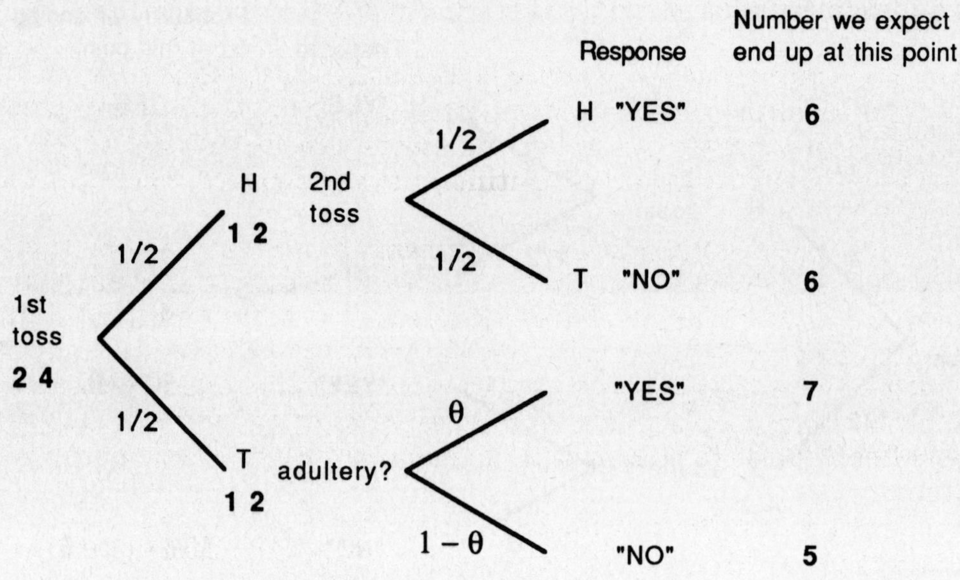


Of course, the estimate of 58% is based on a small sample, and the randomized response technique effectively cuts the sample size in half[3], so the estimate is subject to a large standard error. In this case, the standard error is .203. That is, the true value could easily be anywhere in the range (.381,.787) and could well be anywhere in the interval (.178,.990). This last interval, the estimate ± twice the standard error, is, of course, very wide because the sample size was so small. Nonetheless, the randomized response technique allows one to make an unbiased estimate, while preserving the anonymity of the respondents.

## Exercise

1. Suppose we follow the randomized response protocol with a sample of 40 teachers and find that 17 answer "yes", so that $\hat{p} = 17/40$. What is the estimate of $\theta$?

---

[3] Actually, the standard error of the estimate of $\theta$ is $2*\sqrt{\hat{p}(1-\hat{p})/n}$, which is larger than the standard error one would get using a direct sampling method and a sample of size n/2. In this example the standard error is $2*\sqrt{.542(1-.542)/24} = 2*.097 = .194$, so the estimate that 58% have committed adultery could easily be off by 20% or so. The standard error is large because the sample size is small.

controlled experiment. Rather, surgeons selected patients to be treated, while control patients were chosen at random from a case record index in the same hospital. This is known as using *historical controls*.

In 1979 Creagan, et al.[4], conducted studies of the effectiveness of Vitamin C in treating terminal cancer patients. Theirs was a randomized, controlled, double-blind study: after patients had been admitted to the study, they were randomly assigned to either the treatment group or the control group. Those in the treatment group were given 10 grams per day of vitamin C, while those in the control group were given comparably flavored capsules containing a placebo. The investigators found no differences between the treatment and control groups. Another study, in 1985, confirmed that vitamin C is not effective[5].

The last example warns against the use of historical controls for comparison purposes in a clinical trial. But should all studies be randomized, controlled, and double-blind? No - sometimes, as with most surgical procedures, it is not possible for the trial to be double-blind. Using a control group raises ethical issues, as well. In testing a new treatment for AIDS, it would be unethical to randomly assign half of the patients in the study to a control group and to give these patients a placebo, since we know that these patients would die as a result[6].

---

[4] E. T. Creagan, et al. (1979), "Failure of high-dose vitamin C (ascorbic acid) therapy to benefit patients with advanced cancer," *New Eng. J. Med*, **301**:687-90.

[5] C. G. Moertel, et al. (1985), "High-dose vitamin C versus placebo in the treatment of patients with advanced cancer who have had no prior chemotherapy," *New Eng. J. Med*, **312**:137-41.

[6] For a statistical discussion of ethics and clinical trials for another illness, see J.H. Ware (1989), "Investigating Therapies of Potentially Great Benefit: ECMO," *Stat. Science*, **4**:298-340.

7.4 Paired data

Students who major in mathematics tend to have very strong math skills, but how much stronger are their math skills than their verbal skills? One way to address this question is to examine SAT math (SATM) scores and SAT verbal (SATV) scores for mathematics majors. In a sample of 25 mathematics majors at Oberlin College, the average SATM was 740, while the average SATV was 630.4.

We could compare the SATM and SATV distributions using side-by-side boxplots, but this ignores the fact that the data are *paired*: we have an SATM score and an SATV score for each of 25 students. There is considerable variability between the students. We can control for this variability by computing the difference between SATM and SATV for each student:

$$D = SATM - SATV$$

A boxplot of the differences, D, is given below.

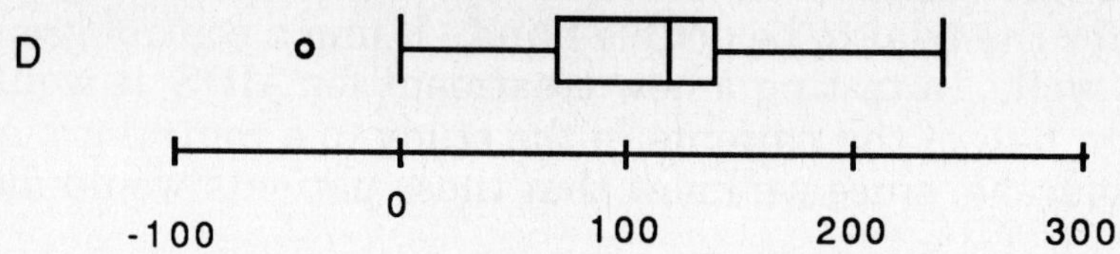


A useful way to explore paired data is with a *sliding square plot*[1]. Such a plot (given below) starts with a scatter plot of SATM vs. SATV and a $45^\circ$ reference line. The SATM scores are projected to the right (to the east), where the marginal distribution (the distribution of just the SATM scores) is shown with a boxplot; likewise, a boxplot of the SATV scores is given at the top of the plot. Finally, points are projected to the southwest (parallel to the reference line) to show the distribution of the differences.

In this plot, we see that most Oberlin mathematics majors score higher on the SATM than on the SATV (most of the points are above the $45^\circ$ reference line). Moreover, the one student with a higher SATV than SATM is an outlier in the distribution of D = SATM - SATV.

---

1 See P. R. Rosenbaum (1989), "Exploratory plots for paired data," Amer. Statist., **43**:108-9.

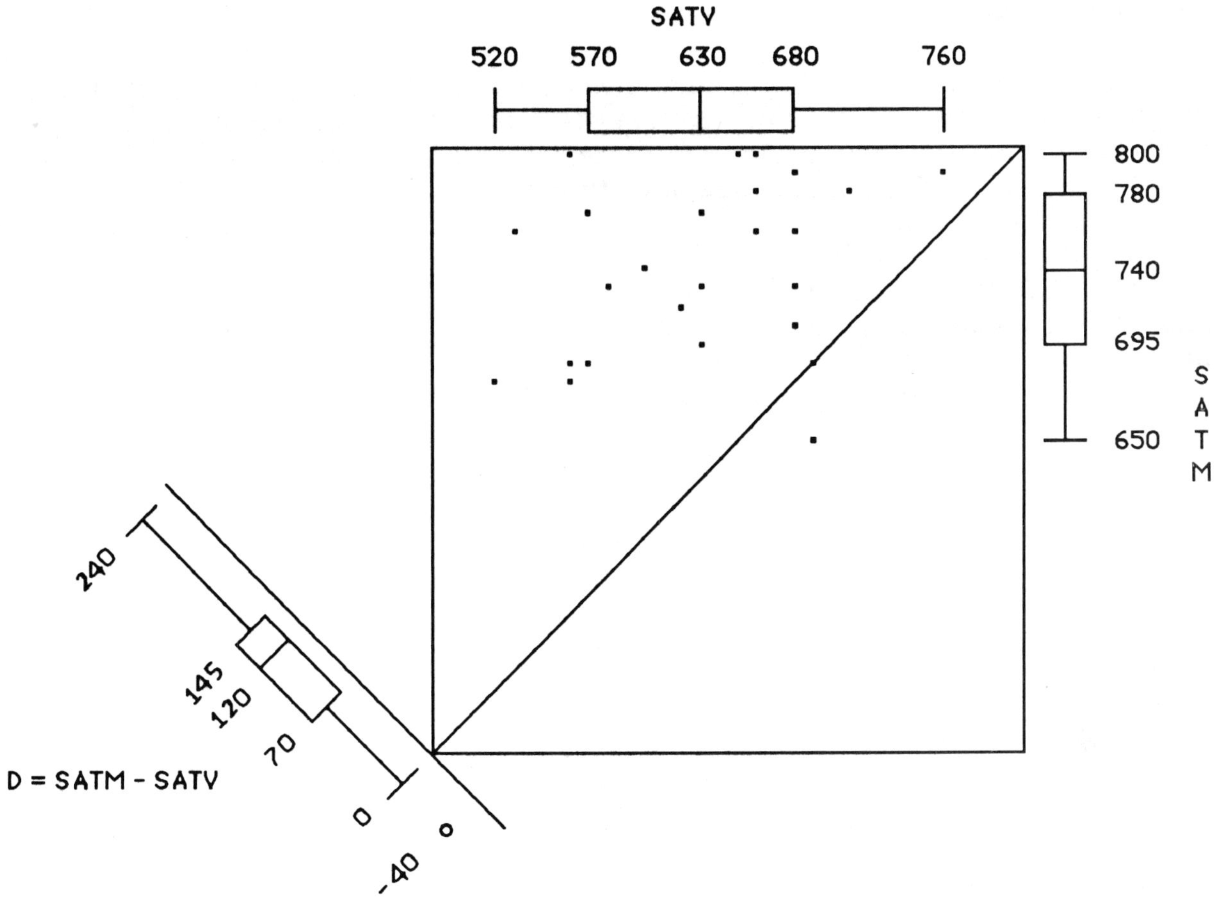


The idea of pairing observations can be extended in a number of ways. Indeed, this idea is of central importance in the statistical design of experiments. One slight variation of the notion of pairing used in the SAT example is *matching*: rather than measure two characteristics of the same subject, one can compare two subjects that are matched (paired) in some way.

For example, in a study to compare two computer programs as teaching aids, we could match students on grade point average by ranking them by GPA and forming a pair out of the first two, a second pair out of the next two, etc. We would then randomly assign one member of each pair to the first group and the other to the second group. At the end of the study we would find the difference in performance of those in the first pair, the difference for those in the second pair, and so on, and would evaluate these differences. This type of pairing is useful when there is a trait, such as scholastic performance (as measured by GPA) that varies greatly from one person to another within the population. Forming pairs by matching subjects on this trait allows us to remove its effect.

## 8.1 Capture/Recapture

Many people are interested in knowing the populations of certain types of wildlife, such as the number of bald eagles in the U.S. or the number of alligators in Florida. There are several ways to estimate such populations[1]; one of the most widely used methods is called capture/recapture.

**Example 1 - Bluegills in Muskellunge Lake**

Naturalists wanted to estimate the number of bluegills in Muskellunge Lake, Indiana[2]. They captured 140 bluegills, marked them, and released them back into the lake. Later they captured a second sample of 727 bluegills and observed that 28 of these had been marked.

Let N denote the number of bluegills in the lake (the quantity to be estimated). If we assume that the proportion of bluegills in the second sample that are marked (the sample percentage) equals the proportion of bluegills in the entire lake that are marked (the population percentage), we can estimate N as follows:

sample percentage marked = population percentage marked

$$\frac{28}{727} = \frac{140}{N}$$

$$N = \frac{140(727)}{28}$$

$$= 3635$$

In general, if M is the number of marked fished in the first sample, n is the number captured in the second sample, and R is the number in the second sample that have been recaptured (i.e., those in the second sample that were marked in the first sample), then we estimate the population, N, as

$$N = \frac{M^*n}{R}$$

There are several ways to mark animals in a capture/recapture experiment. Sometimes a tag is placed on the animal; sometimes fish are marked by removing one of their fins[3].

The capture/recapture method allows one to estimate the population without trying to capture all, or even a large fraction, of it. However, the estimate of N rests upon some assumptions:

(1) The mortality rate among marked animals is the same as that for unmarked animals, and new animals do not enter the population the study.

---

[1] For example, see D. G. Chapman (1989), "The plight of the whales," in J. M. Tanur, et al. (eds.), *Statistics: A Guide to the Unknown*, 3rd edition, Belmont, CA: Wadsworth, pp. 60-67.

[2] W. E. Ricker (1975), *Computation and Interpretation of Biological Statistics of Fish Populations*, Bulletin of the Fisheries Research Board of Canada 191, Ottawa, Canada, pp. 83-6.

[3] Ricker, *op. cit.*

(2) Marked animals do not lose their marks.
(3) In each sample, every animal has the same chance of being captured. In particular, this means that marked animals are not less likely than others to be caught in the second sample, nor is it the case that some animals are "trap happy" (very easy to capture) while others are "trap shy."
(4) The marked animals become randomly mixed with the others between the first and second captures.

The capture/recapture method is not restricted to use in estimating wildlife populations. Estimating the number of homeless people in a large city is difficult, but the capture/recapture method can be used. (Of course, in this case one would not actually put a tag on those captured in the first phase.)

The Census Bureau has developed methods that can be used in adjusting population figures obtained in the decennial census; much of this work is based on the capture/recapture idea[4]. In particular, it is widely held that the census undercounts the populations of minority groups. However, using sampling techniques to adjust the census is controversial. The Census Bureau developed methods of measuring the undercount for the 1990 census, but in 1987 the Department of Commerce decided against making an adjustment[5].

[4] See E. P. Erickson and J. B. Kadane (1985), "Estimating the population in a census year: 1980 and beyond" (with discussion), *J. Amer. Statist. Assoc.*, **80**:98-131.

[5] B. A. Bailar (1988), "Who counts in America?" *Chance*, **1**:9,17.

**Example 1 - The Death Penalty**

In 1978, Warren McCleskey, a black man, was convicted of killing a white police officer and was sentenced to death in Georgia[1]. In an appeal before the U.S. Supreme Court, lawyers for McCleskey argued that the imposition of the death penalty in Georgia was racially biased. They presented statistical models showing that defendants accused of murdering whites in Georgia were four times more likely to be sentenced to death than were defendants accused of murdering blacks and argued that race-of-the-victim discrimination played a key role in the decision to sentence McCleskey to death.

The Supreme Court, in a 5-4 decision, allowed McCleskey's sentence to stand. A full discussion of the legal and statistical issues involved in this case would require hundreds of pages of text. Indeed, many of these issues are the subject of continuing debate[2]. The purpose of this presentation is to examine the major statistical issue of the case and to consider a widespread phenomenon known as *Simpson's paradox*.

There is a pattern in the Georgia data that can be found in the death penalty data from several states. I will focus on data from Florida that make the point clearly[3]. These data, taken from a *Stanford Law Review* article by Samuel Gross and Robert Mauro (Gross and Mauro, 1984), cover homicides in Florida during 1976-1980 in which the victim did not know the suspect. There were 724 such cases in which the suspect was either white or black and the victim was either white or black (cases involving other racial groups are excluded from this analysis). Of these cases, 71 (9.8%) resulted in the suspect being sentenced to death. Table 9.1 provides a breakdown by race of the suspect.

**Table 9.1**

| | | Death Sentence? | | | |
|---|---|---|---|---|---|
| | | Yes | No | Total | % Yes |
| Race of | White | 39 | 308 | 347 | 11.2% |
| Suspect | Black | 32 | 345 | 377 | 8.5% |
| | Total | 71 | 653 | 724 | 9.8% |

Note that white suspects are more likely to be sentenced to death (11.2%) than are black suspects (8.5%).

The racial factor that has the greater effect on the disposition of these cases is not the suspect's race, but the victim's race. Consider a breakdown of the cases in Table 9.1 into two parts: cases involving white victims and cases involving black victims. Of the 724 cases in Table 9.1, 468 involved white victims. In 318 of these

---

[1] McCleskey v. Kemp, 481 U.S. 279, 107 S. Ct. 1756, 95 L. Ed. 2d 262 (1987).

[2] See, for example, D. C. Baldus, C. Pulaski, and G. Woodworth (1983), "Comparative Review of Death Sentences: An Empirical Study of the Georgia Experience," *The Journal of Criminal Law and Criminology*, **74**: 661-753 and R.L. Kennedy (1988), "McCleskey v. Kemp: Race, Capital Punishment, and the Supreme Court," *Harvard Law Review*, **101**: 1388-1443.

[3] These data are taken from S.R.Gross and R. Mauro (1984), "Patterns of Death: An Analysis of Racial Disparities in Capital Sentencing and Homicide Victimization," *Stanford Law Review* **37**:27-153.

cases the suspect was white, while in 150 cases the suspect was black. Black suspects were more likely to be sentenced to death than were white suspects when the victim was white - see Table 9.2.

**Table 9.2**
**White Victims**

| | | Death Sentence? Yes | No | Total | % Yes |
|---|---|---|---|---|---|
| Race of Suspect | White | 39 | 279 | 318 | 12.3% |
| | Black | 29 | 121 | 150 | 19.3% |
| | Total | 68 | 400 | 468 | 14.5% |

Now consider the 256 cases involving black victims. These are summarized in Table 9.3. Black suspects were more likely to be sentenced to death than were white suspects when the victim was black.

**Table 9.3**
**Black Victims**

| | | Death Sentence? Yes | No | Total | % Yes |
|---|---|---|---|---|---|
| Race of Suspect | White | 0 | 29 | 29 | 0.0% |
| | Black | 3 | 224 | 227 | 1.3% |
| | Total | 3 | 253 | 256 | 1.2% |

Comparing these three tables we see an example of Simpson's paradox: blacks are more likely (than whites) to be sentenced to death when the victim is white and blacks are more likely to be sentenced to death when the victim is black. These are the only two settings present and, thus, we might think that blacks would be more likely to be sentenced to death overall (combining the two settings). However, this is not so. Table 9.1 shows that blacks are less likely to be sentenced to death in the aggregate. (Note that combining the numbers in Tables 9.2 and 9.3 yields Table 9.1 - no cases have been added or deleted.)

Compare the "% Yes" values in the tables. In Tables 9.2 and 9.3, the percentage is higher for black suspects than for white suspects, but this inequality is reversed when the data are combined into Table 9.1! How can this be?

The two "% Yes" values for black suspects are 19.3% and 1.3%, which are greater than the "% Yes" values for white suspects of 12.3% and 0%. If we took simple averages the aggregate percentage for blacks would exceed that for whites. But we do not take simple averages. Most of the black suspects were accused of killing blacks, while the overwhelming majority of white suspects were accused of killing whites. Moreover, the death penalty was imposed far more often when the victim was white than when the victim was black. These two factors - (1) whites tend to murder whites while blacks tend to murder blacks and (2) cases with white victims are more likely to result in a death sentence than are cases with blacks victims - combine to produce Simpson's paradox.

The following graph is a pictorial representation of the situation. In the graph, circles represent black suspects and squares represent white suspects, with sizes that are proportional to the number of cases in each category. The aggregate percentage of black suspects sentenced to death is located on the line that connects the two circles and is pulled toward the larger circle. For white suspects the aggregate is pulled toward the larger square.

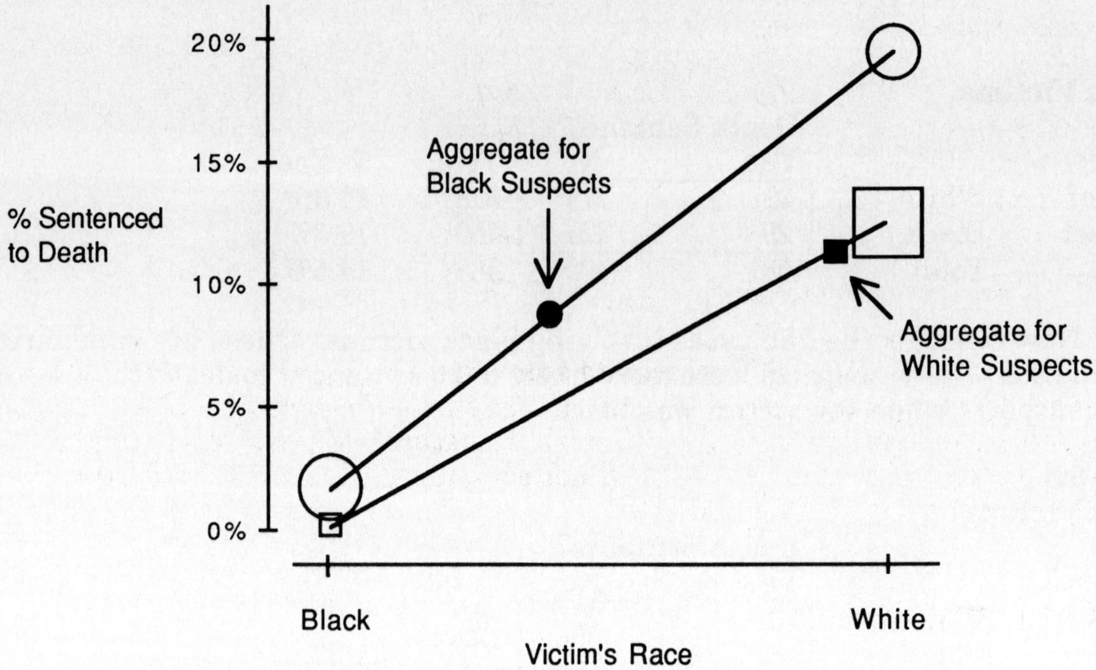


We would say that white suspects are more likely to be sentenced to death than are black suspects <u>if we ignore victim's race</u>, but blacks are more likely to be sentenced to death <u>if we control for victim's race</u>.

Simpson's paradox arises in other situations, as well. For example, Table 9.4 shows average SAT mathematics scores for 1987 and 1988, broken down by ethnic group. Notice that the scores went up in every ethnic category, but the national average did not change. This is due to the fact that students from lower-scoring ethnic groups were a larger percentage of the national total in 1988 than in 1987. Although their scores increased, their larger numbers held the national average steady. Thus, ignoring ethnic groups, average SAT mathematics scores did not change, but controlling for ethnic groups, average scores increased. (Of course, the data in Table 9.4 raise a host of questions regarding racial and cultural biases in standardized tests, equality of educational opportunity, etc.)

**Table 9.4**
**SAT Mathematics Averages**

| | 1987 | 1988 | Change |
|---|---|---|---|
| American Indian | 432 | 435 | +3 |
| Asian | 521 | 522 | +1 |
| Black | 377 | 384 | +7 |
| Mexican-American | 424 | 428 | +4 |
| Puerto Rican | 400 | 402 | +2 |
| Other Hispanic | 432 | 433 | +1 |
| White | 489 | 490 | +1 |
| Other | 455 | 460 | +5 |
| All | 476 | 476 | 0 |

Similarly, average ACT composite scores went down between 1987 and 1988 among students who completed a core curriculum (at least four years of English and at least three years each of mathematics, natural sciences, and social sciences) and among students who did not complete a core curriculum, but the overall average went up - see Table 9.5.

**Table 9.5**
**ACT Composite Averages**

| | | 1987 | 1988 | Change |
|---|---|---|---|---|
| Completed Core | Yes | 21.2 | 21.1 | -0.1 |
| Curriculum? | No | 17.3 | 17.1 | -0.2 |
| | All | 18.7 | 18.8 | +0.1 |

Likewise, it is possible to reduce the income tax rate in every tax bracket while increasing total income tax revenues. As inflation pushes more and more persons into the higher tax brackets, revenues go up even as tax rates go down. (Congress has already figured out how this works.)

## Appendix

The following tables contain the raw data for many of the data sets discussed earlier in the text. (Some sources are given in parentheses.)

Table A-1 1987 High School graduation rates (percentages). (U.S. Dept. of Education.)

| | | | | | |
|---|---|---|---|---|---|
| Alabama | 70.2 | Kentucky | 67.4 | North Dakota | 88.4 |
| Alaska | 66.7 | Louisiana | 60.1 | Ohio | 82.8 |
| Arizona | 64.4 | Maine | 79.3 | Oklahoma | 72.6 |
| Arkansas | 77.5 | Maryland | 74.5 | Oregon | 72.8 |
| California | 66.1 | Massachusetts | 76.5 | Pennsylvania | 78.7 |
| Colorado | 73.7 | Michigan | 62.4 | Rhode Island | 69.4 |
| Connecticut | 80.5 | Minnesota | 90.6 | South Carolina | 66.9 |
| D.C. | 55.5 | Mississippi | 64.8 | South Dakota | 79.7 |
| Delaware | 70.1 | Missouri | 74.4 | Tennessee | 67.8 |
| Florida | 58.6 | Montana | 86.2 | Texas | 65.1 |
| Georgia | 62.5 | Nebraska | 86.7 | Utah | 80.6 |
| Hawaii | 70.8 | Nevada | 72.1 | Vermont | 78.0 |
| Idaho | 78.8 | New Hampshire | 72.7 | Virginia | 74.0 |
| Illinois | 75.7 | New Jersey | 77.2 | Washington | 77.8 |
| Indiana | 73.7 | New Mexico | 71.7 | West Virginia | 76.2 |
| Iowa | 86.4 | New York | 62.9 | Wisconsin | 85.4 |
| Kansas | 82.1 | North Carolina | 67.8 | Wyoming | 89.3 |

Table A-2 June rainfall (inches) for Cleveland, OH. (Climatological Data, Ohio, and Local Climatological Data, Cleveland, Ohio; National Oceanic and Atmospheric Administration, U.S. Dept. of Commerce.)

| | | | | | |
|---|---|---|---|---|---|
| 1938 | 2.45 | 1952 | 1.48 | 1966 | 1.83 |
| 1939 | 3.25 | 1953 | 3.30 | 1967 | 1.17 |
| 1940 | 3.68 | 1954 | 1.38 | 1968 | 2.32 |
| 1941 | 2.79 | 1955 | 2.68 | 1969 | 4.61 |
| 1942 | 4.09 | 1956 | 4.14 | 1970 | 4.98 |
| 1943 | 4.02 | 1957 | 5.85 | 1971 | 3.79 |
| 1944 | 3.04 | 1958 | 4.67 | 1972 | 9.06 |
| 1945 | 4.41 | 1959 | 1.82 | 1973 | 6.72 |
| 1946 | 5.14 | 1960 | 3.01 | 1974 | 3.57 |
| 1947 | 6.07 | 1961 | 2.95 | 1975 | 4.10 |
| 1948 | 3.95 | 1962 | 2.95 | 1976 | 3.64 |
| 1949 | 1.63 | 1963 | 1.93 | 1977 | 4.91 |
| 1950 | 3.77 | 1964 | 2.06 | 1978 | 3.30 |
| 1951 | 3.25 | 1965 | 3.05 | | |

Table A-3 Los Angeles Dodger home attendance, 4/13/89-7/3/89.

| Weekends | Weekdays |
|---|---|
| 44312 | 47136 |
| 43836 | 32489 |
| 39520 | 33619 |
| 47251 | 48295 |
| 48437 | 27630 |
| 48493 | 29310 |
| 46767 | 34861 |
| 47389 | 36689 |
| 46329 | 32199 |
| 47034 | 37089 |
| 47083 | 27781 |
| 46060 | 44953 |
| 46912 | 25336 |
| 42207 | 22858 |
| 41032 | 28161 |
| 33372 | 29073 |
| 42940 | 30430 |
| 43050 | 42782 |
| 46411 | 37385 |
| 33692 | 33708 |
| 40789 | 36128 |
| | 29673 |

Table A-4 1952-1988 Presidential election participation - % voting. (Federal Election Commission.)

| | |
|---|---|
| 1952 | 61.6 |
| 1956 | 59.3 |
| 1960 | 62.8 |
| 1964 | 61.9 |
| 1968 | 60.9 |
| 1972 | 55.2 |
| 1976 | 53.5 |
| 1980 | 54.0 |
| 1984 | 53.3 |
| 1988 | 50.1 |

Table A-5 1973-1977 U.S. household larcenies, by month, in thousands. (National Crime Survey Report, U.S. Dept. of Justice.)

| | | | | | |
|---|---|---|---|---|---|
| 511 | Jan. 1973 | 753 | | 803 | |
| 408 | | 760 | | 941 | |
| 479 | | 683 | | 967 | July 1976 |
| 598 | | 717 | | 929 | |
| 674 | | 575 | Jan. 1975 | 748 | |
| 740 | | 573 | | 755 | |
| 769 | July 1973 | 668 | | 742 | |
| 812 | | 679 | | 749 | |
| 634 | | 778 | | 602 | Jan. 1977 |
| 664 | | 865 | | 574 | |
| 614 | | 1011 | July 1975 | 669 | |
| 635 | | 933 | | 758 | |
| 614 | Jan. 1974 | 825 | | 831 | |
| 698 | | 813 | | 921 | |
| 663 | | 697 | | 984 | July 1977 |
| 711 | | 807 | | 1002 | |
| 707 | | 643 | Jan. 1976 | 789 | |
| 845 | | 638 | | 821 | |
| 865 | July 1974 | 673 | | 701 | |
| 917 | | 713 | | 766 | |

Table A-6 1934-1987 Cleveland average annual temperature. (Climatological Data, Ohio, and Local Climatological Data, Cleveland, Ohio; National Oceanic and Atmospheric Administration, U.S. Dept. of Commerce.)

| | | | | | |
|---|---|---|---|---|---|
| 1934 | 50.9 | 1952 | 52.1 | 1970 | 49.9 |
| 1935 | 50.1 | 1953 | 53.0 | 1971 | 49.8 |
| 1936 | 50.3 | 1954 | 52.3 | 1972 | 48.5 |
| 1937 | 50.4 | 1955 | 52.3 | 1973 | 52.6 |
| 1938 | 52.7 | 1956 | 50.9 | 1974 | 50.1 |
| 1939 | 52.3 | 1957 | 51.0 | 1975 | 50.5 |
| 1940 | 49.0 | 1958 | 49.4 | 1976 | 48.6 |
| 1941 | 51.8 | 1959 | 51.8 | 1977 | 49.3 |
| 1942 | 50.7 | 1960 | 48.1 | 1978 | 49.2 |
| 1943 | 49.6 | 1961 | 49.7 | 1979 | 49.2 |
| 1944 | 51.1 | 1962 | 49.0 | 1980 | 48.0 |
| 1945 | 50.0 | 1963 | 47.4 | 1981 | 49.1 |
| 1946 | 52.5 | 1964 | 49.8 | 1982 | 50.1 |
| 1947 | 50.2 | 1965 | 49.9 | 1983 | 51.0 |
| 1948 | 50.7 | 1966 | 48.5 | 1984 | 49.0 |
| 1949 | 52.9 | 1967 | 49.7 | 1985 | 49.4 |
| 1950 | 49.7 | 1968 | 48.6 | 1986 | 50.8 |
| 1951 | 50.5 | 1969 | 49.0 | 1987 | 51.4 |

Table A-7 Male mourning dove "nest materials gathering times", in seconds - see Section 3.2. (These are personal observations made on 8/5/89.)

| | | | | |
|---|---|---|---|---|
| 90 | 205 | 130 | 52 | 69 |
| 67 | 58 | 35 | 38 | 45 |
| 48 | 38 | 64 | 34 | 108 |
| 37 | 87 | 23 | 72 | 45 |
| 33 | 49 | 59 | 28 | 29 |
| 40 | 66 | 24 | 72 | 30 |
| 77 | 104 | 149 | 34 | 56 |
| 71 | 34 | 115 | 25 | 23 |
| 61 | 112 | 168 | 31 | 65 |
| 34 | 68 | 81 | 22 | 28 |
| 33 | 42 | 70 | 54 | 51 |
| 31 | 46 | 37 | 32 | 58 |
| 79 | 56 | 27 | 74 | 55 |
| 30 | 29 | 143 | 102 | 42 |
| 83 | 68 | 74 | | |

Table A-8 State demographic data, 1980 census. (State and Metropolitan Data Book 1986, U.S. Dept. of Commerce.)

| State | Per capita income | % college graduates | % high school graduates | median age | % over age 65 |
|---|---|---|---|---|---|
| Alabama | 7454 | 12.6 | 56.5 | 29.3 | 11.3 |
| Alaska | 13007 | 22.4 | 82.5 | 26.1 | 2.9 |
| Arizona | 8855 | 16.8 | 72.4 | 29.2 | 11.3 |
| Arkansas | 7099 | 9.7 | 55.5 | 30.6 | 13.7 |
| California | 11020 | 19.8 | 73.5 | 29.9 | 10.2 |
| Colorado | 10147 | 23 | 78.6 | 28.6 | 8.6 |
| Connecticut | 11559 | 21.2 | 70.3 | 32 | 11.7 |
| D.C. | 12210 | 28.1 | 67.1 | 31.1 | 11.6 |
| Delaware | 9949 | 16.3 | 68.6 | 29.8 | 10 |
| Florida | 9245 | 14.7 | 66.7 | 34.7 | 17.3 |
| Georgia | 8041 | 15.3 | 56.4 | 28.7 | 9.5 |
| Hawaii | 10129 | 20.3 | 73.8 | 28.4 | 7.9 |
| Idaho | 8100 | 16.1 | 73.7 | 27.6 | 9.9 |
| Illinois | 10448 | 14.5 | 66.5 | 29.9 | 11 |
| Indiana | 8970 | 12.4 | 66.4 | 29.2 | 10.7 |
| Iowa | 9211 | 14.1 | 71.5 | 30 | 13.3 |
| Kansas | 9799 | 15.7 | 73.3 | 30.1 | 13 |
| Kentucky | 7644 | 11 | 53.1 | 29.1 | 11.2 |
| Louisiana | 8404 | 13.4 | 57.7 | 27.4 | 9.6 |
| Maine | 7756 | 14 | 68.7 | 30.4 | 12.5 |
| Maryland | 10365 | 19.8 | 67.4 | 30.3 | 9.4 |
| Massachusetts | 10096 | 20 | 72.2 | 31.2 | 12.7 |
| Michigan | 9798 | 15.2 | 68 | 28.8 | 9.8 |
| Minnesota | 9662 | 16.7 | 73.1 | 29.2 | 11.8 |

| | | | | | |
|---|---|---|---|---|---|
| Mississippi | 6614 | 13 | 54.8 | 27.7 | 11.5 |
| Missouri | 8856 | 14 | 63.5 | 30.9 | 13.2 |
| Montana | 8334 | 17.3 | 74.4 | 29 | 10.7 |
| Nebraska | 8887 | 16.1 | 73.4 | 29.7 | 13.1 |
| Nevada | 10845 | 15.1 | 75.5 | 30.2 | 8.2 |
| New Hampshire | 9217 | 18.4 | 72.3 | 30.1 | 11.2 |
| New Jersey | 10811 | 18.6 | 67.4 | 32.2 | 11.7 |
| New Mexico | 7947 | 17.3 | 68.9 | 27.4 | 8.9 |
| New York | 10242 | 18.7 | 66.3 | 31.9 | 12.3 |
| North Carolina | 7774 | 13.4 | 54.8 | 29.6 | 10.3 |
| North Dakota | 8651 | 15.2 | 66.4 | 28.3 | 12.3 |
| Ohio | 9401 | 14.8 | 67 | 29.9 | 10.8 |
| Oklahoma | 9029 | 15.7 | 66 | 30.1 | 12.4 |
| Oregon | 9319 | 17.2 | 75.6 | 30.2 | 11.5 |
| Pennsylvania | 9352 | 13.8 | 64.7 | 32.1 | 12.9 |
| Rhode Island | 9200 | 15.3 | 61.1 | 31.8 | 13.4 |
| South Carolina | 7389 | 14.2 | 53.7 | 28.1 | 9.2 |
| South Dakota | 7815 | 14.2 | 67.9 | 28.9 | 13.2 |
| Tennessee | 7689 | 11.9 | 56.2 | 30.1 | 11.3 |
| Texas | 9439 | 16 | 62.6 | 28.2 | 9.6 |
| Utah | 7679 | 20.3 | 80 | 24.2 | 7.5 |
| Vermont | 7970 | 19.5 | 71 | 29.4 | 11.4 |
| Virginia | 9446 | 19.2 | 62.4 | 29.8 | 9.5 |
| Washington | 10248 | 18.8 | 77.6 | 29.8 | 10.4 |
| West Virginia | 7747 | 10.5 | 56 | 30.4 | 12.2 |
| Wisconsin | 9361 | 14.9 | 69.6 | 29.4 | 12 |
| Wyoming | 11009 | 17.2 | 77.9 | 27.1 | 7.9 |

Table A-9 Minnesota Twins 1989 batting averages.

| Player | Average before All-Star game | Average after All-Star game |
|---|---|---|
| Puckett | 0.341 | 0.336 |
| Dwyer | 0.292 | 0.327 |
| Harper | 0.333 | 0.313 |
| Gladden | 0.298 | 0.293 |
| Castillo | 0.198 | 0.292 |
| Moses | 0.278 | 0.283 |
| Hrbek | 0.267 | 0.280 |
| Gaetti | 0.216 | 0.268 |
| Bush | 0.267 | 0.260 |
| Gagne | 0.295 | 0.254 |
| Backman | 0.178 | 0.254 |
| Newman | 0.255 | 0.251 |
| Larkin | 0.303 | 0.235 |
| Laudner | 0.259 | 0.201 |

Table A-10 Brain weights and body weights for various species - see Section 5.2.

| species | brain wt (gm) | body wt (kg) |
|---|---|---|
| canary | 0.848 | 0.0171 |
| crow | 9.3 | 0.337 |
| flamingo | 8.05 | 1.598 |
| loon | 6.12 | 1.53 |
| ostrich | 42.11 | 123 |
| pheasant | 3.289 | 0.625 |
| pigeon | 2.694 | 0.282 |
| stork | 16.24 | 3.35 |
| vulture | 19.6 | 5.27 |
| catfish | 1.84 | 2.894 |
| barracuda | 3.83 | 5.978 |
| mackerel | 0.64 | 0.765 |
| salmon | 1.257 | 3.93 |
| brown trout | 0.57 | 0.292 |
| tuna | 3.09 | 5.21 |
| northern trout | 1.233 | 2.5 |
| grizzly bear | 233.9 | 142.88 |
| cheetah | 2.449 | 22.2 |
| lion | 106.7 | 28.79 |
| raccoon | 40 | 5.175 |
| skunk | 10.3 | 1.7 |
| tiger | 302 | 209 |
| wolf | 152 | 29.94 |
| greyhound | 105.9 | 24.49 |
| seal | 442 | 107.3 |
| walrus | 1126 | 667 |
| porpoise | 1735 | 142.43 |
| blue whale | 6800 | 58059 |
| bat | 0.936 | 0.028 |
| mole | 1.16 | 0.0396 |
| baboon | 140 | 7.9 |
| grey monkey | 66.6 | 4.55 |
| chimpanzee | 440 | 56.69 |
| human | 1377 | 74 |
| mouse | 0.551 | 0.0177 |
| squirrel | 3.97 | 0.183 |
| rhinoceros | 655 | 763 |
| African elephant | 5712 | 6654 |
| horse | 618 | 461.76 |

Table A-11 1830-1990 population of Florida (U.S. Census Bureau).

| | |
|---|---|
| 1830 | 34730 |
| 1840 | 54477 |
| 1850 | 87445 |
| 1860 | 140424 |
| 1870 | 187748 |
| 1880 | 269493 |
| 1890 | 391422 |
| 1900 | 528542 |
| 1910 | 752619 |
| 1920 | 968470 |
| 1930 | 1468211 |
| 1940 | 1897414 |
| 1950 | 2771305 |
| 1960 | 4951560 |
| 1970 | 6791418 |
| 1980 | 9747197 |
| 1990 | 12937926 |

Table A-12 Massachusetts burglary data for 1988. (Uniform Crime Reports of the F.B.I., U.S. Dept. of Justice.)

| City | 1988 Burglaries | Population |
|---|---|---|
| Boston | 9163 | 579921 |
| Brockton | 1913 | 94741 |
| Fall River | 1997 | 89595 |
| Malden | 446 | 53892 |
| Medford | 349 | 57256 |
| New Bedford | 2352 | 95570 |
| Quincy | 787 | 83864 |
| Somerville | 537 | 72822 |
| Springfield | 3694 | 149842 |
| Waltham | 561 | 57518 |
| Weymouth | 385 | 55333 |
| Worcester | 3378 | 159124 |

## Index

ACT scores 113
adultery 100
airplane landing data 45
analysis of covariance 86
Anscombe 74

bar chart 17
baseball attendance 15, 115
baseball salaries 7
basketball salaries 9
batting averages 5, 67, 118
bias
  nonresponse 97
  panel 103
  selection 97
bluegill example 108
body weight data 49, 71
brain weight
  and body weight data 79, 119
boxplots 11
  side-by-side 13
burglary data 80, 120

capture/recapture 108
census 109
cereal data 87
Challenger accident 46, 53, 65
class rating data 72
clock response time data 94
coin data 38
college degrees 18
college credits data 44
common cold 103
compound smoothers 30
confounding 103
Cook's distance 81
correlation 51

death penalty 110
divorce rates 5
dotplots 6
draft lottery 21
dummy variable 87

experiments 103
  blind 104
  double-blind 104
exponential smoother 31

fidelity 98
football winning margin 36
french fries 1, 6

Gould 86
graph of averages 64

heights of students 13
high school graduation rates 2, 6, 11, 114
highway fatality rate data 92
histogram 6
historical controls 105
home runs 4, 37

income 62, 70, 84, 117
influential points 80
interaction plot 94
interactions 92
interquartile range 12

larceny data 25, 30, 116
laughter and funniness data 95
least-squares 62
library gate count data 57, 76
library science budget data 71, 78
line chart 19
lineplots 6
Literary Digest 97

M&Ms 17
magazine advertisements 9
mammary artery ligation 104
matching 107
median 3
median age by state 3, 81
median fit line - see resistant line
median trace 21
mourning doves 41, 117
multiple regression 84

National Newspaper Poll 98
normal probability plot 32
normal probability plots
  residuals 76
normal quantile plot 33
normal scores 33

observational studies 103
outlier 3, 12, 80

paired data 106
pie chart 18
placebo effect 103
power transformation 39
presidential election 13, 15, 19, 115
presidents
    age at inauguration 5

Q-Q plot 33
quartile 11

$r^2$ 66, 86
racial distributions 18
rainfall - see weather
randomized response 100
range 12
regression 62, 68
    effect 67
    fallacy 68
residual 62
residual plots 70
residuals 31
resistant lines 57
running medians 27

SAT
    math majors data 106
    national averages 112
scatter plot 46
scatterplot matrix 88
sentence length 42
Simpson's paradox 110
sliding square plot 106
smoothers 25
snowfall -see weather
stem-leaf diagram 1
    back-to-back 3
surveys 97
swimming times 32

temperature - see weather
time series 25
transformations 38, 77
treatments 103
Tufte, E. 20

vitamin C and cancer 104

water use 26
weather
    rainfall 13, 37, 114
    snowfall 1, 5, 61
    temperature 48, 53, 61, 116